AF836576

Formulation and Device Lifecycle Management of Biotherapeutics

Formulation and Device Lifecycle Management of Biotherapeutics

A Guidance for Researchers and Drug Developers

BEATE BITTNER
Global Product Strategy, Product Optimization,
F. Hoffmann-La Roche Ltd., Basel, Switzerland

JOHANNES SCHMIDT
Global Product Strategy, Product Optimization,
F. Hoffmann-La Roche Ltd., Basel, Switzerland

ELSEVIER

ACADEMIC PRESS
An imprint of Elsevier

Academic Press is an imprint of Elsevier
125 London Wall, London EC2Y 5AS, United Kingdom
525 B Street, Suite 1650, San Diego, CA 92101, United States
50 Hampshire Street, 5th Floor, Cambridge, MA 02139, United States
The Boulevard, Langford Lane, Kidlington, Oxford OX5 1GB, United Kingdom

Copyright © 2022 Elsevier Inc. All rights reserved.

No part of this publication may be reproduced or transmitted in any form or by any means, electronic or mechanical, including photocopying, recording, or any information storage and retrieval system, without permission in writing from the publisher. Details on how to seek permission, further information about the Publisher's permissions policies and our arrangements with organizations such as the Copyright Clearance Center and the Copyright Licensing Agency, can be found at our website: www.elsevier.com/permissions.

This book and the individual contributions contained in it are protected under copyright by the Publisher (other than as may be noted herein).

Notices
Knowledge and best practice in this field are constantly changing. As new research and experience broaden our understanding, changes in research methods, professional practices, or medical treatment may become necessary.

Practitioners and researchers must always rely on their own experience and knowledge in evaluating and using any information, methods, compounds, or experiments described herein. In using such information or methods they should be mindful of their own safety and the safety of others, including parties for whom they have a professional responsibility.

To the fullest extent of the law, neither the Publisher nor the authors, contributors, or editors, assume any liability for any injury and/or damage to persons or property as a matter of products liability, negligence or otherwise, or from any use or operation of any methods, products, instructions, or ideas contained in the material herein.

British Library Cataloguing-in-Publication Data
A catalogue record for this book is available from the British Library

Library of Congress Cataloging-in-Publication Data
A catalog record for this book is available from the Library of Congress

ISBN: 978-0-12-823741-0

For Information on all Academic Press publications
visit our website at https://www.elsevier.com/books-and-journals

Publisher: Andre G. Wolff
Acquisitions Editor: Erin Hill-Parks
Editorial Project Manager: Susan Ikeda
Production Project Manager: Swapna Srinivasan
Cover Designer: Christian J. Bilbow

Typeset by MPS Limited, Chennai, India

Working together
to grow libraries in
developing countries
www.elsevier.com • www.bookaid.org

Contents

Introduction

Today, the pressure on healthcare costs and resources is increasing, and especially for biopharmaceuticals that require parenteral administration, the inherent complex and invasive dosing procedure adds to the demand for efficient medical management. In light of the COVID-19 pandemic the value of drug delivery technologies in enabling a flexible care setting is broadly recognized. In such a setting, patients and their caregivers can choose the place of drug administration based on individual preferences and capabilities. This includes not only dosing in the clinic but also supervised at-home dosing, and self-administration for eligible patients.

Formulation and device lifecycle management is a multifunctional discipline in the pharmaceutical industry that covers all aspects of improving the way a medicine is administered to the patient. Activities focus on optimizing a molecule's drug delivery profile, such as the administration route or the dosing regimen, as well as aspects of the product presentation. The ultimate aim is to reduce the dosing complexity and, thus, increase patient convenience and minimize drug administration-related costs and capacity bottlenecks in the healthcare system. Formulation and device lifecycle management can start at any stage of a molecule's lifecycle, but the term is most commonly used to describe activities following initial approval of a medicine. In the literature, synonyms for "formulation and device lifecycle management" could be "product optimizations" or "drug delivery improvements."

The purpose of this book is to provide an overview on various aspects of formulation and device lifecycle management for biotherapeutics, including guidance on when and how to conduct related development activities. The target audience are multidisciplinary researchers and drug developers in pharmaceutical industry, biotech companies, and academia involved in the formulation and device development. Beside scientists covering technical aspects of product optimizations, researchers and drug development experts in nonclinical, clinical, access, and commercial functions are presumably interested in the content of the book.

Previously, the mission of drug delivery scientists pertaining to parenteral administration of biotherapeutics was to enable drug administration per se. Especially for high-dose monoclonal antibodies, this typically involved the development of an intravenous formulation. Products were predominantly dosed based on an individual patient's body size in analogy to cytotoxic agents with a narrow therapeutic window. Nowadays, product optimizations for biological medicines are conducted to improve patient experience with the dosing regimen and to address healthcare-related societal needs.

From a drug delivery perspective, this includes simplification of drug administration via the development of subcutaneous formulations or injection devices with connected features that facilitate contact between patients and their healthcare provider in a remote setting. Efforts are ongoing to ensure that dosing can be broadly done with fixed doses instead of body weight- or body surface area-adjusted dosing regimens. In addition, fixed-dose combinations that contain two or more biological molecules within the same formulation are developed to complement parenteral combination therapy. For other administration routes, such as ocular and brain delivery, research on long-acting formulation and implant technologies is ongoing striving for patient centricity via less invasive and less frequent drug administration schedules.

Chapter 1 provides a broad review of the present status of formulation and device lifecycle management for biotherapeutics with the focus on product optimizations for already approved biological medicines. Subsequently, in Chapter 2, the scientific basis and nonclinical and clinical development pathway for subcutaneous dosing alternatives to established intravenous treatments for high-dose monoclonal antibodies in oncology is described. These subcutaneous dosing alternatives that currently still require administration in a controlled healthcare environment result in markedly reduced hospital resource utilization associated with the parenteral dosing procedure. To realize the dosing of suitable biotherapeutics outside of the clinic, availability of ready-to-use subcutaneous injection devices is mandated. Chapter 3 summarizes the current requirements for validating autoinjector or wearable on-body delivery systems clinically and describes a concept on how to further streamline pharmacokinetic comparability studies for device platforms that can be used for several products across different indications.

While subcutaneous dosing in the home setting is already a reality in disease areas such as rheumatoid arthritis, multiple sclerosis, or primary immunodeficiency, especially in cancer care, biological drugs are to a great extent still dosed in a hospital, physician's office, or infusion center. Chapter 4 discusses the opportunities and barriers to enabling a flexible care setting in oncology based on apparent patient and healthcare provider needs. To define the next steps for realizing home- and self-administration of eligible biological anticancer agents that have been shown to be safe and well tolerated, the chapter includes a comprehensive precedent assessment of biological drugs from other disease areas that are registered for patient self-administration. Special focus is on the molecules' drug delivery profile and the product presentation.

Chapter 5 illustrates multidisciplinary development and commercialization aspects that are typically transferable between different molecules and aspects that need to be assessed on a case-by-case basis.

Chapter 6 aims at guiding nonclinical and clinical drug delivery scientists as well as professionals in market access and commercialization roles on the initiation of formulation and device optimizations during the lifecycle of a biological medicine. The chapter provides a decision architecture built on scientific, technical, and health economic information that is becoming available at the different 5 development stages, allowing manufacturers to make informed decisions on when to best embark into specific development activities at reduced investment risk.

It is recommended to start reading chapters 1 to 4 that provide detailed insights into different aspects and underlying concepts of formulation and device lifecycle management, followed by chapters 5 and 6 which take up the initial learnings and provide specific guidance to drug developers and researchers in the field. The individual sections of the book are however written to also be suitable as stand-alone readings, by briefly summarizing prior information relevant to understand the context without the need to read all previous chapters.

Product optimizations are to a large extent molecule independent and manufacturers can therefore leverage learnings from drugs that applied the same formulation or device technology for other molecules in different indications.

The authors are grateful to all researchers and drug development scientists involved in formulation and device lifecycle management. Their ongoing efforts helped facilitate biological dosing regimens, increase access to medications, and generate a wide-ranging understanding on the relevance of the field in enabling truly patient-centric drug delivery.

The book was written during the COVID-19 pandemic in 2020. At this time, parenteral drug delivery outside of the hospital and closer to or in the patient's home, especially in cancer care, shifted from a nice-to-have to a necessity. It is assumed that this remarkable change in the perception of the value of drug delivery will further accelerate innovations in the field.

CHAPTER 1

Enabling customer-centric and sustainable drug delivery—formulation and device lifecycle management of biotherapeutics

Introduction

The term "formulation and device lifecycle management" covers all activities during a medicine's development and commercialization period that aim at improving the way the molecule is administered to a patient. Associated product optimizations typically reduce dosing complexity, improve patient experience with the treatment regimen, lessen drug delivery-related healthcare costs and resources, and thus differentiate a product on the market. Inherent to their physical–chemical properties, especially biotherapeutics that need to be administered parenterally are expected to benefit from improved drug delivery solutions (Bittner et al., 2018).

To guarantee earliest access to new potentially lifesaving medications, time to market often needs to be prioritized over development of a more convenient product presentation. Formulation and device lifecycle management that minimizes dosing complexity then becomes highly relevant to address the increasing pressure on the global healthcare system and to broaden access to medications. Product optimizations leading to the simplified administration of biotherapeutics can improve dosing accuracy and adherence through better patient experience, which can in turn result in an improved treatment effect (Menditto et al., 2020). Associated lifecycle management activities include changes in route of administration, that is, from intravenous to subcutaneous to oral, the use of fixed doses instead of body weight- or body surface area-adjusted dosing, less frequent dosing regimens, fixed-dose combinations (FDCs) with two or more molecules coformulated in the same dosing vehicle, and ready-to-use and connected drug delivery devices.

Cancer immunotherapy is a disease area with various novel biotherapeutics on the market where companies early on introduced a number of product optimizations. For example, pembrolizumab (Keytruda) was introduced in 2014 as

Formulation and Device Lifecycle Management of Biotherapeutics
DOI: https://doi.org/10.1016/B978-0-12-823741-0.00007-5
© 2022 Elsevier Inc.
All rights reserved.

an intravenous treatment with body weight-adjusted dosing. As part of lifecycle management, the dosing principle has been changed to a fixed dose in 2016 (Freshwater et al., 2017), and currently development of a subcutaneous formulation is ongoing (ClinicalTrials.gov, 2018c). Similar optimizations are being done for the cancer immunotherapy treatments, atezolizumab (Tecentriq) and nivolumab (Opdivo) (ClinicalTrials.gov, 2018b; Felip et al., 2021).

One of the commercially most successful examples of formulation and device lifecycle management is adalimumab (Humira), initially launched in 2003 (Azevedo et al., 2016). The manufacturer introduced a considerable number of product improvements including packaging and storage conditions, dosage modifications, and device enhancements (Norman, 2016). Also, a number of biosimilars start formulation and device lifecycle management early on. For example, Truxima, a rituximab biosimilar, improved the storage conditions of the formulation soon after initial launch in 2016 (Kim et al., 2019). Remsima, infliximab's biosimilar is available with both an intravenous and a subcutaneous formulation, while branded infliximab (Remicade) can only be dosed intravenously (Reinisch et al., 2019).

The aim of this chapter is to describe examples of formulation and device lifecycle management for the administration of biotherapeutics across different therapeutic areas. Focus is on technologies for biopharmaceuticals that (1) improve the logistics around supply chain, as well as drug preparation and administration in a healthcare institution, (2) reduce the treatment burden via reduced dosing complexity and time expenditure in a healthcare institution, and possibly enable a flexible care setting, and (3) minimize dosing complexity inherent to combination therapy with two or more molecules (Fig. 1.1). In scope are product optimizations for drugs that are already on the market with a different, more complex administration schedule.

The optimal timing of exploring drug delivery improvements and the role of applying platform formulation or device technologies, as well as the impact of the healthcare professional reimbursement model on the successful adaptation of a novel formulation or device, will be discussed.

Impact of formulation and device lifecycle management opportunities for biotherapeutics

Improved product presentation

Optimizations in a medicinal product's shelf-life, long-term or bench stability, and packaging configuration are an established, integral part of formulation and device lifecycle management. Here, seemingly small

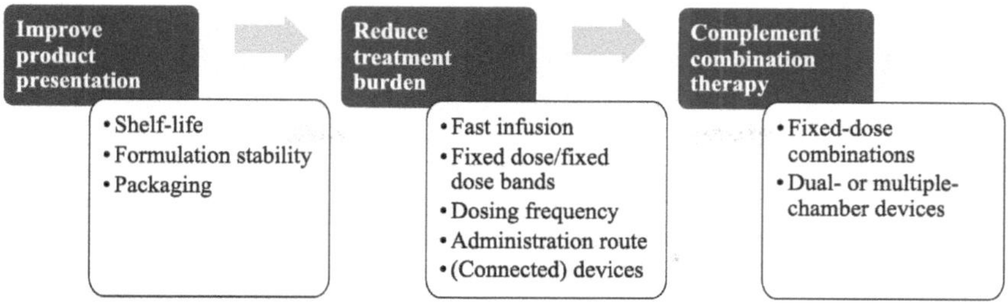

Figure 1.1 Formulation and device lifecycle management for administration of biotherapeutics. Overview of formulation and device lifecycle management activities that (1) improve the logistics around supply chain, and drug preparation and administration in a healthcare institution, (2) reduce the treatment burden via reduced dosing complexity and time expenditure in a healthcare institution, and enabling a flexible care setting, and (3) minimize dosing complexity inherent to combination therapy of two or more biopharmaceuticals.

changes can play a pivotal role in simplifying processes in the clinic, hospital pharmacy, or the patient's home, as well as in reducing the overall supply chain complexity.

Shelf-life and stability of the formulation

Shelf-life is the period of time, from the date of manufacture, that a drug product is expected to remain within its approved product specification while stored under defined conditions. A short shelf-life of biopharmaceutical formulations paired with the need for controlled cool storage and shipment imposes severe constraints on the inventory management and increases the supply chain complexity (Edelman, 2004; Graves et al., 2009).

Inherent to the limited long-term stability of many biopharmaceuticals in liquid high-concentration solutions (Garidel et al., 2017), freeze-dried lyophilizates that need to be reconstituted prior to administration are frequently applied formulation alternatives with an increased shelf-life (Haeuser et al., 2020). Such drug presentations cannot be administered directly to patients, as they are not presented in ready-to-administer liquid form. Reconstitution in hospital pharmacies is prone to a number of errors (Suvikas-Peltonen et al., 2017) that could be avoided by having a liquid dosing vehicle. Typical risks related to poor reconstitution practices in hospital pharmacies include calculation errors leading to administration of the wrong dose, incorrect reconstitution (insufficient mixing, incomplete dissolution, use of the wrong diluent), incompatibility between diluent, infusion, other medicinal products or administration devices, or the risk of microbiological contamination (Scheepers et al., 2015).

The availability of a liquid formulation is therefore considered advantageous if it proves to be sufficiently stable throughout the desired shelf-life of ideally at least 2 years to be economically viable (Bye et al., 2014). Today, there is a higher percentage of approved liquid parenteral protein formulations (66%) compared to lyophilized formulations (34%) (Gervasi et al., 2018). To enable fast patient access to novel treatment modalities, initial launch with a lyophilizate, followed by a ready-to-use liquid formulation, should therefore be an integral part of lifecycle management for biopharmaceuticals (Copaxone, 2014).

The stability of a biological drug at ambient temperature is an additional important factor to be considered. Short stability can result in drug wastage and increased cost, for example, in case a patient arrives late to the appointment. While this issue can in parts be overcome by on-demand preparations in which the nurse would contact the pharmacy when the next dose is needed, the safety of such ad hoc preparation and dispensing can be affected by competing priorities, interruptions, and distractions in the pharmacy environment (Burger & Massoomi, 2019).

Packaging

In addition to a medication's formulation, the packaging configuration plays an important role in its safe and cost-effective application. Packaging is about protection and security, and thus, reducing any potential errors due to mistaken medication identity is a priority for hospital pharmacies. Some hospitals will purposely separate products that look too similar to avoid errors (Shah & Shah, 2011). Hospital pharmacies that require smaller quantities of a given medication may prefer single-dose packages demanding less storage space and are easier to return. Multidose packages can be an advantage for high volume items but often require repackaging prior to use.

Packages should be designed to avoid counterfeit drugs to the extent possible (Bansal et al., 2013). This includes unit-dose coding and use of serialization, a comprehensive track-and-trace system allowing identification of every product package based on the assigned serialization number. An ideal anticounterfeit technology should be secure and noncloneable. The technology should be difficult to remove and reapply, easy to check, automatically authenticated, be useable by consumers, and must be legally compliant by the industries (Bansal et al., 2013).

Reduced treatment burden

In addition to the above improvements in a biopharmaceutical's formulation and packaging that mainly simplify drug preparation processes and reduce related costs and resources in the clinic and hospital pharmacy, further optimizations in drug delivery can result in a significant reduction of the overall treatment burden. Namely, fast infusion intravenous dosing regimens, less frequent administration schedules, fixed doses or fixed-dose bands instead of body size-adjusted dosing enabling the use of injection devices, or a change in the administration route can greatly raise patient experience and reduce the pressure on the overall healthcare system.

Fast infusion

Infusion-related reactions are defined as any signs or symptoms experienced by patients during the infusion of pharmacological or biological agents or any event occurring on the first day of drug administration. These reactions predominantly occur during the initial infusion(s) and can be characterized by flushing, rash, fever, rigors, chills, dyspnea, mild, and/or severe hypotension with(out) bronchospasms, cardiac dysfunction, and/or anaphylaxis (Doessegger & Banholzer, 2015; Kang & Saif, 2007).

In clinical practice, one way to reduce the rate and extent of hypersensitivity and infusion-related reactions is to continue dosing at a slower intravenous infusion rate (Chung, 2008; Lenz, 2007). In addition, patients are typically observed postinfusion for several hours during the first treatment cycle(s) in the infusion clinic, which impacts center capacity and increase waiting time for subsequent patients (Rees et al., 2020).

In a systematic review on infusion reactions to and infusion rate of monoclonal antibodies (mAbs) used in cancer treatment, Rombouts et al. (2020) found that for some mAbs, the incidence of infusion reactions is low or can be lowered by the use of adequate premedication schedules. Moreover, dependent on the individual mAb, faster infusion schedules are possible. Reassessing the required infusion duration based on individual patient's tolerability and increasing knowledge on the overall safety profile of a mAb is an important measure to enable shorter infusion times as a lifecycle management opportunity. Such time savings are not only expected to improve patient experience with the treatment but also benefit outpatient infusion facilities by increasing both infusion chair capacity available during clinical hours and nursing efficiencies (Can et al., 2013).

Fixed dose or fixed dose bands

Historically, many biotherapeutics were dosed on a body size-adjusted basis (Egorin, 2003). Disadvantages of body weight- or body surface area-adjusted dosing regimens include the risk of dose calculation errors or of product wastage, the relative complexity of dose preparation, and the difficulties to develop cost-effective ready-to-use injection devices, such as autoinjectors (AIs) or wearable on-body delivery systems (OBDSs).

Consequently, today, the feasibility of developing fixed doses or dose bands with one dose for all patients, independent of their body weight or body surface area, or based on different weight ranges, respectively, is already assessed during early clinical development (Quartino et al., 2017; Stroh et al., 2017). For biologics that were initially approved with a body size-adjusted regimen, fixed doses can be introduced as a lifecycle management product optimization. For example, the immune checkpoint inhibitors nivolumab and pembrolizumab added an intravenous fixed-dose alternative (Freshwater et al., 2017; Zhao et al., 2017) to their established weight-based dosing regimen. The IL-6 inhibitor tocilizumab and trastuzumab in HER2-positive breast cancer introduced a novel subcutaneous fixed-dose regimen (Fettner et al., 2019; Wynne et al., 2013) that supported development of their AI and OBDS presentations.

The feasibility of fixed-dose regimens is supported by the observation that mAbs typically distribute within the blood plasma and extracellular fluids only, which increase less than proportionally with an increase in body weight. Depending on the mAb, intracellular degradation after binding to the target can be the primary route of elimination and is not affected by body size (Hendrikx et al., 2017). Moreover, analyzing pharmacokinetic and pharmacodynamic data from mAb studies published in peer-reviewed journals, Wang et al. (2009) compared the impact of body size-based dosing and fixed dosing on PK variability in adults and showed that both approaches perform similarly.

Change in dosing frequency

Today, biopharmaceuticals are predominantly administered parenterally (i.e., intravenously, subcutaneously, or intramuscularly) and depending on the administration volume and dosing complexity, drug administration frequently takes place in a healthcare institution. Recurring parenteral administrations can therefore be burdensome for patients and healthcare providers. Inherent to their comparatively long half-lives (Ovacik & Lin, 2018), dosing frequencies of biologics, such as mAbs, are typically in the range of once weekly to once every 6 months. Despite these moderate to long treatment intervals, dosing a high number of patients can still be challenging for infusion clinics with capacity constraints. Consequently, efforts made to decrease dosing frequency are typically valued. Especially when it comes to combination therapy with two or more biopharmaceutical molecules (Henricks et al., 2015), harmonization of the dosing frequency plays a pivotal role in reducing the treatment load both for patients and the healthcare system.

While less frequent parenteral dosing regimens are generally preferred, there are still patients in favor of more frequent administration (Augustovski et al., 2013; Williams & Edwards, 2006), possibly related to staying in regular contact with their treating physician or to receiving lower individual doses. Moreover, patient preference assessments reveal that individual patients are more likely to fill the prescription and adhere to the therapy when their preferred dosing regimen is available, so that the availability of more than one regimen for the same molecule can be beneficial (Cowan et al., 2019).

A number of technology platforms aiming at reducing the dosing frequency of a subcutaneously administered biological are already on the market or in development. Such platforms aim at either (1) enlarging the dosing volume that can be administered to a single injection site, (2) increasing the concentration of the biological molecule in the dosing solution, or (3) sustaining the release of an injected molecule into the systemic circulation.

Alternative administration route

The development of alternative administration routes as a lifecycle management pursuit for biopharmaceuticals, such as mAbs, aims at reducing invasiveness and increasing patient experience with the treatment regimen. Ideally, shifting drug administration to the patient's home and thus allowing for flexibility in site of care are a key focus area for dosing route changes of biopharmaceuticals.

Subcutaneous delivery

A major improvement in parenteral delivery of mAbs is the availability of subcutaneous instead of or as an alternative to intravenous infusions. Following injection into the subcutaneous tissue, absorption of biopharmaceuticals is comparatively slow and incomplete due to presystemic catabolism. A major reason for the slow absorption is the transport to the blood or lymph capillaries in the subcutis (Richter & Jacobsen, 2014).

Subcutaneous administration is broadly preferred over intravenous dosing by patients and healthcare providers and has been demonstrated to reduce administration-related cost and resources. This administration route is an established procedure for products requiring doses of less than 200 mg per injection, which in most cases can be formulated in volumes of 1–2 mL. Such relatively small volumes can be injected using currently available prefilled pen (PFP) and AI devices (Bittner & Schmidt, 2021).

Biological therapies with dosing volumes exceeding 2 mL are outside of the standard AI or PFP delivery range, yet can still be delivered via a slow subcutaneous infusion either manually or with a suitable syringe pump. Alternatively, a ready-to-use wearable OBDSs might be developed to further simplify administration. The first biological medicine that is available with a larger volume device is evolocumab (Repatha), a mAb that inhibits proprotein convertase subtilisin/kexin type 9. Evolocumab is marketed in combination with a 3.5-mL device, Repatha Pushtronex, copackaged with a prefilled cartridge to deliver 420 mg of active within 9 minutes. The high-volume device provides a monthly dosing alternative instead of three subcutaneous injections of 140 mg provided in a prefilled syringe (PFS) or in a single-use prefilled AI, SureClick (Boccara et al., 2017; Stadler & Cook, 2017).

A complementary approach to facilitate subcutaneous administration of high-dose biotherapeutics is liquid high-concentration formulation technology platforms that minimize the administration volume. Such approaches aim at generating stable high-concentration solutions within a viscosity range that allows syringeability and technical processing and are based on a number of different technology concepts. EXCELSE, for example, is a high-concentration technology using a customized blend of amino acids that coat key regions of the biological to stabilize and to prevent aggregation, thus enabling concentrations of up to 250 mg/mL. Arsia Therapeutics, acquired by Eagle Pharmaceuticals in 2016 (Eagle Pharmaceuticals, 2016), makes use of viscosity-lowering agents, such as camphor sulfonic acid mixtures with arginine or lysine, supporting mAb concentrations up to approximately 250 mg/mL depending on the biological molecule (Larson et al., 2015).

Another approach is the development of stable suspensions that may have the potential to achieve concentrations exceeding 400 mg/mL according to their manufacturers. Technology examples are XeriJect with water free paste formulations to allow subcutaneous administration with available PFPs or pumps (Newswanger et al., 2015; Xeris Pharmaceuticals, 2020) or Elektroject with microparticle-based suspension formulations (ONdrugDelivery, 2019).

Microglassification uses a similar approach by removing the majority of water from biological solutions leading to stable solid, spherical, and amorphous microbeads that could be injected through a 27-G needle (Microglassification, 2019). All of these technologies are still in early-stage development and applicability to a broader number of biopharmaceuticals, long-term stability, and clinical acceptability still remain to be demonstrated.

For rapid injection of higher volumes exceeding 5–10 mL use of the permeation and absorption enhancing technology platform recombinant human hyaluronidase PH20 (rHuPH20) has been shown to facilitate subcutaneous infusion at all injection sites (e.g., abdomen, thigh, or arm) by spreading the drug over a wider area within the subcutaneous tissue (Frost, 2007). An approved biological using rHuPH20 is HyQvia, a subcutaneous immunoglobulin for the treatment of primary immunodeficiency and hypogammaglobulinemia. HyQvia is marketed as a dual vial unit containing 10% IgG and rHuPH20 for sequential administration. The subcutaneous infusion is initiated with rHuPH20 followed within 10 minutes by the subcutaneous immunoglobulin through the same needle (Hyqvia, 2014). The first marketed subcutaneous mAbs that utilize a coformulation with rHuPH20 were trastuzumab (Herceptin, Herceptin Hylecta) and rituximab (MabThera, Rituxan Hycela) for the treatment of HER2-positive breast cancer and lymphoma, respectively (Shpilberg & Jackisch, 2013).

Sustained release formulation technology platforms represent an additional approach to optimize subcutaneous administration by extending the dosing intervals and improving tolerability due to lower peak exposures of the drug (Škalko-Basnet, 2014). These platforms include biodegradable polymers, such as poly(lactic-co-glycolic acid) (PLGA) microspheres or hydrogels exemplified by cross-linked hyaluronic acid or polycaprolactonedimethacrylate and hydroxyethyl methacrylate that can be loaded with therapeutic proteins. Successful hydrogel or biodegradable polymer platforms currently release the active via different mechanisms including diffusion, hydrogel degradation, enzymatic cleavage, or combinations thereof.

Oral delivery

Oral delivery is generally considered the most preferred administration route (Renukuntla et al., 2013) as it provides several advantages over injection, including lack of invasiveness, greater simplicity of administration, increased patient convenience and compliance, avoidance of injection site pain, discomfort, or infections. With these advantages the oral administration route enables home- and patient self-administration and typically has lower manufacturing costs associated with the lack of need for sterile production (Barclay et al., 2013; Borner et al., 2002; Dibonaventura et al., 2010; Fasano, 1998; Gupta et al., 2013; Hamman et al., 2005; Park et al., 2011; Renukuntla et al., 2013; Yun et al., 2013).

Oral delivery of biotherapeutics is challenged by their low bioavailability (typically less than 1%) due to rapid degradation in the gastrointestinal tract and their poor permeability across the intestinal mucosa. Consequently, strategies to overcome a poor oral bioavailability would need to protect the molecule from pH-dependent and enzymatic degradation in the digestive tube and facilitate absorption into the systemic circulation (Philippart et al., 2016; Vllasaliu et al., 2018).

One of the most advanced oral delivery platforms is the Eligen technology, salcaprozate sodium (SNAC) (Twarog et al., 2019). SNAC has been applied in the development of the first oral glucagon-like peptide-1 receptor agonist semaglutide (RYBELSUS) approved in 2017 (Rybelsus, 2017). SNAC increases passive transcellular permeation across the small intestinal epithelia based on noncovalent macromolecule complexation and is associated with a pH-elevating, monomer-inducing, and pepsin-inhibiting effect in the stomach (Twarog et al., 2019). Oral semaglutide tablets are available for once-a-day intake and are an alternative to weekly injections with its subcutaneous product presentation (Ozempic, 2017).

An advanced example of an oral biopharmaceutical in clinical development is octreotide administered with an oily suspension formulation. In this dosing vehicle, octreotide is solubilized with sodium caprylate and polyvinyl pyrrolidone in water, milled, lyophilized, and dispersed in a lipophilic medium comprised of polysorbate 80, glyceryl monocaprylate, and glyceryl tricaprylate. Drug/oil suspensions are then administered in enteric-coated capsules (Tuvia et al., 2014). Phase 3 of oral octreotide in acromegaly was a multicenter, open-label, dose-titration, baseline-controlled study that enrolled 155 patients. Almost two-third of patients with complete or partially controlled insulin growth factor-1 and growth hormone levels, and receiving injectable somatostatin receptor ligand (SRL) for 3 months or longer, maintained hormonal control after switching to the oral treatment, with a safety profile consistent with approved SRLs (Melmed et al., 2015).

From the oral technologies that are currently under investigation, an ingestible microneedle-based device, developed to ensure intact passage of an incorporated therapeutic protein through the gastrointestinal tract with subsequent direct insertion into the intestinal tissue from where the biological drug is released and absorbed into systemic circulation, might represent a promising technology (Abramson et al., 2019; Schoellhammer et al., 2016). However, additional research and clinical data are required to confirm the suitability for human use.

Pulmonary delivery

Despite the convenience advantages of subcutaneous over intramuscular and intravenous administration of biopharmaceuticals, psychological barriers associated with injectables, such as needle phobia or difficulties with the injection technique, still remain a barrier for acceptance in the broader population (McLenon & Rogers, 2019). In this context, pulmonary administration might represent an alternative and less invasive administration route that facilitates patient self-administration.

Pulmonary drug delivery enables targeting the drug directly to the lung both for local and systemic treatment (Patil & Sarasija, 2012). Following inhalation, a molecule accesses the highly permeable and large surface area of the alveolar epithelium. The extent of penetration into the lungs and final site of disposition is depending on the size of inhaled drug particles that is a composite of the formulation and aerosolization steps, as well as the breathing patterns, the specific anatomy of an individual's airways, and the disease state (Morales et al., 2017; Patton & Byron, 2007).

To facilitate pulmonary self-administration, different inhalation devices are available. Pressurized metered-dose inhalers (pMDIs) and dry powder inhalers (DPIs) are the two most commonly used devices in the treatment of pulmonary diseases (Da Rocha et al., 2011). pMDIs deliver the medicine in the form of a short burst of aerosolized fluid (Sanders et al., 2017), while transport of the dry powder from a DPI is typically breath activated (Atkins, 2005).

Prominent lifecycle management examples for a change from subcutaneous to pulmonary delivery are two rapid-acting inhaled insulins, marketed as Exubera in 2006 and Afrezza in 2014 (Oleck et al., 2016). Exubera had been developed to provide a more physiological prandial insulin replacement than regular subcutaneous human insulin for patients with diabetes, with the expected advantage of non-invasive delivery (Barnett, 2004). Pulmonary administration of insulin was based on a chemically stabilized protein, a dry powder formulation exhibiting required aerosol physical characteristics, a low-dose powder filling and packaging technology, and a mechanical device for powder dispersal and reliable dosing to the patient (White et al., 2005). However, Exubera was withdrawn from the market already after less than 2 years due to a number of product-related disadvantages compared to subcutaneous insulins and limited acceptance by patients and physicians (Kalra et al., 2019). The most common side effect of Exubera was hypoglycemia and many respiratory adverse effects were reported. Moreover, six newly diagnosed cases of lung malignancies were found in patients taking Exubera (Mitri & Pittas, 2009). Different to subcutaneous insulins that are dosed in units, Exubera was dosed in milligrams (mg), which is a novel regimen from the user perspective. In addition, administering Exubera involved an inhaler system perceived as unwieldy that may have mitigated perceived convenience of inhalation and the fear of injections may also have been overestimated in lifetime injectable users. Another drawback, especially from a payer perspective, was that Exubera was launched at a premium prize in a market where the current standard of care is effective and safe, and prescribers were unfamiliar with the novel administration route. For example, the German Institute for Quality and Efficiency in Health Care saw no evidence that inhaled insulin (Exubera) in diabetes therapy shows advantages over short-acting human insulin or insulin analogs administered subcutaneously (Heinemann, 2008).

While Afrezza's safety profile is similar to that of Exubera (Afrezza, 2014), the second to market inhaled insulin, offers a number of advantages. The associated inhaler is smaller and generally considered more discreet to use and insulin is dosed in unites and not in mg and thus easier to comprehend for patients and prescribers (Oleck et al., 2016). The long-term acceptance of the treatment still remains to be awaited.

In pulmonary drug delivery in general, the availability of an inhaler device that is addressing individual patient needs has to be taken into consideration. An example for a patient-centered device lifecycle management is dornase-alpha (Pulmozyme), a mucolytic agent indicated to reduce the risk of respiratory infections and to improve pulmonary function in patients with cystic fibrosis (Pulmozyme, 2018). Cystic fibrosis is characterized by a significant daily treatment effort, including the burden of nebulized therapies (Hassan et al., 2016). Dornase-alpha is inhaled once- to twice-daily (Pulmozyme, 2018) and involves a time-consuming preparation, inhalation and cleaning procedure. A novel nebulizer system (eRapid) applying a vibrating membrane technology was developed to deliver dornase-alfa more rapidly and efficiently than jet nebulizers without affecting the physicochemical properties of the drug (Scherer et al., 2011).

Administration of dornase-alfa via this nebulizer resulted in comparable efficacy and safety, shorter inhalation times, and higher patient preference compared to established jet nebulizers (Pari LC Plus) (Sawicki et al., 2015). A key advantage of the novel inhaler device was the reduction of the total daily inhalation time from approximately 30 minutes with the jet nebulizers to approximately 10 minutes with the vibrating membrane nebulizer (Naehrig et al., 2011).

Given the convenience advantage of pulmonary drug delivery, as a minimally invasive administration route, using environment- and user-friendly inhalation devices, today, a variety of biopharmaceuticals that were previously developed with a subcutaneous formulation have reached clinical trials studying delivery with an inhaler formulation (Morales et al., 2017).

Ocular delivery

Ocular drug delivery aims at targeting a molecule directly to its place of action, the human eye. The human eye can be partitioned into the anterior and posterior segments. The anterior segment contains the cornea, iris, lens, and ciliary muscle, whereas the posterior segment encompasses aqueous/vitreous humor, choroid, macula, retina and optic nerve. This complex structure of the eye including the presence of static and dynamic barriers in conjunction with the efflux pumps generates a challenge for delivery of the treatment (Gaudana et al., 2010).

Vision impairment can stem from different ophthalmic diseases, including neovascular (wet) age-related macular degeneration (AMD), cataract, diabetic macular edema, diabetic retinopathy, glaucoma, retinal detachment, retinal pigmentation, or uveitis (Mecklenburg & Schraermeyer, 2007; Pascolini & Mariotti, 2012).

Several biologics are approved for the treatment of ophthalmic diseases and drug delivery-related lifecycle management activities are ongoing. Namely, ranibizumab (Lucentis) and aflibercept (Eylea) are extracellular vascular endothelial growth factor inhibitors (anti-VEGF) administered via intravitreal injections (Eylea, 2019; Lucentis, 2017). Intravitreal injection procedures for the administration of anti-VEGF therapy have initially been conducted in an operating room environment. In a number of countries, including Canada and the United States, for cost and effectiveness reasons, the majority of anti-VEGF therapy has moved to an office setting but still remains resource intensive for ophthalmologists and causes a high burden on patients and on the healthcare system (Grzybowski et al., 2018; Lau et al., 2018). Less-frequent dosing therefore addresses a major unmet need exemplified by a survey of 1029 retina specialists rating "reduced treatment burden" and "long acting/sustained delivery" with 66% and 70%, respectively, the greatest unmet needs regarding AMD treatment (Monés et al., 2020). The need for therapies requiring less-frequent intravitreal injections is reflected in real-world clinical practice, demonstrated in the AURA study, a multicountry, retrospective, observational study in real-life anti-VEGF treatment in patients with wAMD. The mean number of intravitreal injections patients received declined in this study from 5.0 in the first year to 2.2 injections in the second year (Holz et al., 2015). A likely factor contributing to the diminishing adherence over time is the treatment burden for patients. A survey in patients on glaucoma treatment revealed increasing preference for intravitreal injections with 26%, 46%, and 82% for monthly, bi-monthly, and six-monthly administration, respectively (Torres et al., 2018).

The convenience advantage of less frequent dosing schedules presents an opportunity for anti-VEGF formulation and device lifecycle management. Development of a sustained release presentation requiring less injections and office visits is expected to improve patient's compliance while reducing costs and injection-related complications of anti-VEGF therapy and is therefore a key focus for product lifecycle management in ophthalmology. A change in the route of administration to topical delivery has not been successful to date. Structural barriers in the eye complicate the transport of drugs. Especially, for biologics hurdles for topical delivery are even higher due to a decrease in scleral permeability with increasing molecular weight (Rodrigues et al., 2018).

Development of sustained delivery formulations and devices to optimize the release profile and dosing frequency appears to be a more promising approach. A number of long-acting anti-VEGF delivery systems for intravitreal injection are currently in development. This includes liposomes consisting of a lipid bilayer built of phospholipids and cholesterol that can incorporate bipolar molecules, such as anti-VEGF, in the inner aqueous compartment from where they are released over a longer period of time (Seah et al., 2020). Another approach for a long-acting release technology in development is dendrimers. Dendrimers consist of polymers with repeating and regularly branched units. Sustained release is generally approached by covalently attaching biomolecules via chemical conjugation, which, however, poses a risk for anti-VEGF therapy due to potential deactivation of the protein (Seah et al., 2020).

Hydrogels are 3-dimensional polymer networks with a high-water content that have the capacity to encapsulate hydrophilic molecules. The cross-linked polymer structure allows preparation of hydrogels to a tunable solid that offers a wide range of properties including incorporation and release of proteins (Li & Mooney, 2016). These properties make hydrogels a promising technology for sustained release systems of biologics. Depending on the hydrogel system drug release over a few days to over a year has been reported (Ashley et al., 2013). For anti-VEGF molecules (aflibercept and ranibizumab), coformulation with PLGA microspheres in polyethylene glycol-poly(L-lactic acid)-diacrylate/poly(*N*-isopropylpropylacrylamide) (PEG-PLLA-DA/pNIPAAm) or with PLGA microspheres suspended within pNIPAAm-PEG-DA hydrogel achieved in vitro release durations of 180–200 days, respectively (Chang et al., 2019). To advance the hydrogel technology especially for intravitreal administration, further development work is needed to deliver a durable drug release over several months and complete clearance of the hydrogel.

The development of implants has become another path for sustained intraocular delivery of biologics. Although controlled release profiles over several months can be achieved, the major challenge is the placement of the implant requiring a surgical procedure. Therefore the requirements for implants with regards to durability, magnitude of sustained delivery as well as the complexity of the surgical procedure need to outweigh the risk and the burden for patients and the healthcare system compared to currently available therapies.

Two principle implant-based drug delivery approaches in development: (1) biodegradable implants predominantly made of PLGA and (2) refillable, nonbiodegradable implants. Biodegradable implants have the advantage that there is no removal required following drug release. There is, however, currently no biodegradable implant containing a biological approved for eye disease available commercially (Iyer et al., 2019). A refillable, nonbiodegradable implant in late stage development with an anti-VEGF molecule is the port delivery system (PDS). The PDS is a refillable reservoir containing a release control element (RCE) and a self-sealing septum. The drug is released into the vitreous through the RCE via passive diffusion. The RCE has been engineered for ranibizumab providing a sustained ranibizumab release exceeding 6 months. In the LADDER phase 2 study, the median time to first implant refill was between 8.7 and 15 months depending on the drug load of the ranibizumab formulation (Campochiaro et al., 2019). Based on these positive results, development of the ranibizumab PDS has been advanced into phase 3 (ClinicalTrials.gov, 2018a). The pivotal study in 418 patients with wAMD evaluated the efficacy and safety of the PDS device filled with 100 mg/mL ranibizumab and refilled every 6 months at fixed intervals compared to monthly intravitreal injections of 0.5 mg ranibizumab (10 mg/mL ranibizumab formulation). Achievement of the primary endpoint of the study, change in best-corrected visual acuity score from baseline at the average of weeks 36 and 40, demonstrated equivalence between monthly intravitreal ranibizumab injections and the PDS refilled every 6 months (Patel et al., 2021).

Availability of a ready-to-use PFS for intravitreal administration represents another ocular lifecycle management strategy. Although use of PFS has no direct impact on the treatment burden for patients, injection by means of a PFS instead of a single-use vial and handheld syringe reduces the preparation time and potentially the risk of endophthalmitis (Sassalos & Paulus, 2019). In a Danish study, the preparation time for intravitreal injections was assessed in 172 cases. The total preparation time starting from a vial was 40.3–45.1 seconds, and using a PFS saved 25.5 seconds compared to the vial (95% confidence interval: 23.3 to 27.6 seconds, $P < .0001$). The amount of time saved by eliminating preparation steps can increase institutional resource utilization. The reduced risk of contaminations with the ready-to-use PFS additionally supports potential outsourcing to non-ophthalmologists (Subhi et al., 2016). Examples of anti-VEGF therapeutics with a PFS presentation introduced as a lifecycle management are aflibercept (Eylea) (Eylea, 2019), pagatinib sodium (Macugen) (Macugen, 2011), and ranibizumab (Lucentis) (Lucentis, 2017).

Brain delivery

Targeted delivery to the brain, especially for biotherapeutics, is still one of the highest hurdles for the development of effective treatments of neurological diseases, including Alzheimer's (Pardridge, 2020), Parkinson's and Huntington's disease (Teixeira et al., 2020) or brain cancers (Jena et al., 2020). The key challenge is to overcome the blood–brain barrier (BBB), a protective barrier between the blood and brain interstitial fluid. This barrier acts as a dynamic conduit allowing the passage of essential nutrients, peptides, proteins, and immune cells into the brain, while hindering the delivery of drugs to the brain. It is estimated that 98% of small molecules and almost 100% of large molecule drugs do not cross the BBB (Pardridge, 2020). This currently hampers development and subsequent lifecycle management of biological drugs for neurological diseases requiring significant drug exposure in the brain (Teixeira et al., 2020).

Novel formulation approaches aiming at effectively and noninvasively targeting biopharmaceuticals to the brain include polymeric nano- or microparticles, lipidic nanoparticles, inorganic nanoparticles, liposomes, nanogels, nanoemulsions, dendrimers, quantum dots, as well as the intranasal drug delivery to the brain (see the "Nasal Delivery" section) (Alexander et al., 2019). Emerging invasive brain delivery technologies include lumbar puncture for intrathecal administration (Hinderer et al., 2020), direct ventricular delivery (DeVos & Miller, 2013), or cisterna magna delivery (Katz et al., 2018).

Transdermal delivery

Molecules delivered through the skin bypass the hepatic first-pass metabolism and, additionally, transdermal delivery could provide a sustained drug release enabling reduction in dosing frequency. These features have generated interest to investigate transdermal delivery as an alternative route of administration for peptides and proteins and thus the availability of a transdermal dosing alternative is expected to increase cost-effectiveness and improve patient compliance compared to more invasive administration routes (Bajracharya et al., 2019).

The main impediment for transdermal absorption of biological drugs is the stratum corneum, a skin barrier protecting the body from the influx of foreign molecules, such as toxins (Barry, 2002). A number of different approaches or combinations thereof to overcome this barrier are in development including drug delivery systems, such as nanoparticles and lipid-based delivery systems, or the use of physical methods, such as electroporation, iontophoresis, sonophoresis, and microneedles (Jain et al., 2017). Despite the many efforts and progress in the development of technologies facilitating transdermal administration first approvals of biological therapies delivered through the skin will be important to drive wider use of the technologies for the lifecycle management of large molecules.

Nasal delivery

The nasal route is obtaining increasing attention as a noninvasive administration route and could become a viable lifecycle management approach for proteins and peptides (Rohrer et al., 2018). Intranasal administration bypasses degradation in the gastrointestinal tract and provides direct access to the cerebrospinal fluid (CSF) within the olfactory region of the brain (Crowe et al., 2018).

This delivery route might therefore represent a more convenient alternative to intrathecal injection of drugs into the CSF in the future.

There are currently a number of biological therapeutics with an intranasally administered dosing alternative on the market. Products include desmopressin acetate (marketed as DDAVP, Minirin, Desmotabs, Desmospray, Octim, and others, depending on the formulation and country of license) provided as nasal spray for the treatment of nocturnal enuresis and central cranial diabetes insipidus (Hashim & Abrams, 2008). In a randomized, double-blind, double-dummy, cross-over study in 30 pediatric patients with nocturnal enuresis, nasal delivery of desmopressin was compared with an oral tablet formulation. The study demonstrated comparable efficacy of an intranasal dose of 20 mg compared to an oral dose of 200 mg desmopressin (Fjellestad-Paulsen et al., 1987). This observed 10-fold lower dose required using the intranasal administration may be an important factor when designing cost-efficient lifecycle management strategies for biological therapeutics.

Another lifecycle management example is the nasally administered vaccine FluMist Quadrivalent (FluMist Quadrivalent, 2019) provided in a prefilled nasal sprayer encompassing a single 0.2 mL dose to be administered as 0.1 mL per nostril for patients of 9 years or older. The intranasal spray is an alternative to Fluzone Quadrivalent (Fluzone Quadrivalent, 2020) requiring intramuscular injection of 0.5 mL for the same patient population. Both vaccines are indicated for active immunization for the prevention of influenza disease caused by influenza A subtype viruses and type B viruses contained in the vaccine.

Limitations of intranasal administration include the low dosing volume of typically less than 200 μL due to the restrained tissue surface area, low permeability for hydrophilic drugs, short retention time, and the need for coformulation of absorption enhancers (Erdő et al., 2018).

Connected devices and accompanying apps

In settings where parenteral biotherapeutics are administered at home, with the help of a nonprofessional caregiver or by the patient her- or himself, adherence can be reduced as compared to drug administration in healthcare institutions. Underlying causes can be patients' beliefs about the necessity of medicines, difficulties in understanding the administration procedure, undesirable side effects, or simply forgetfulness (Santus et al., 2019; Solsona et al., 2017).

To further facilitate drug delivery and ensure long-term adherence and persistence, an emerging product optimization in self-administration areas is the development of so-called connected devices and accompanying health apps. Connected features would typically consist of educational tools, dosing reminders, adherence trackers, and the option for patients to record adverse events or other disease-related observations in an electronic diary. A first step when developing such electronic tools is to identify individual barriers to medication adherence among the respective patient population. These barriers then need to be addressed via population- or individual patient-specific measures. Interventions need to consider patient and caregiver capabilities, should challenge dysfunctional beliefs, and support the development of routines or habits (Arden et al., 2019).

Today, especially inhalation devices are equipped with adherence trackers and dosing reminders but are typically applied for small molecule formulations. Examples for connected devices for the subcutaneous administration of biotherapeutics are the SMARTInhaler and the Propeller medication sensors used for a variety of differed inhaler types (Chan et al., 2015; Merchant et al., 2018). The RebiSmart (Lugaresi, 2013) and the Betaconnect (Ziemssen et al., 2015) were the first subcutaneous injection devices complementing home-administration of interferons in multiple sclerosis (MS). The first connected PFP for insulin, the InPen, with an associated smartphone app makes dosing recommendations and prepares reports for healthcare professionals (Gildon, 2018; Klonoff & Kerr, 2018).

Drug delivery to complement combination therapy

As a result of the increasing number of biotherapeutics developed for combination therapy (Henricks et al., 2015), manufacturers are mandated to implement technologies that avoid a further increase in dosing complexity to the extent possible. In addition to aligning dosing frequency for two or more parenterally administered molecules and a change from more invasive intravenous to subcutaneous administration, the development of subcutaneous FDCs and dual- or multiple-chamber injection devices is in the focus of current research in the field.

Fixed-dose combinations

FDCs are defined as the coformulation of two or more active molecules in the same dosing vehicle. Such coformulations with small molecules have been a reality in disease areas with polypharmacy for many years and their purposes include improving patient compliance, simplifying the dosing schedule, reducing risk of adverse events, enabling cheaper shipment, and reduced packaging quantities (Ugurlu & Ozaydin, 2014). It is expected that these advantages will be even more pronounced for biopharmaceuticals, where a simplification of a parenteral dosing regimen will have a larger impact as compared to reducing the number of oral dosage forms for small molecules that need to be taken during the day. Also, typically package sizes for biopharmaceuticals are larger and products need to be stored under cool conditions, so that having one instead of two or more packages will largely facilitate shipment and storage requirements.

Looking at the landscape for FDCs containing two or more mAbs, the majority of these coformulations are developed with an intravenous dosing regimen. The most advanced FDC development is the subcutaneous coformulation of pertuzumab and trastuzumab (Phesgo) that has been approved by the FDA (Phesgo, 2020) based on the phase 3 registration study FeDeriCa demonstrating noninferior pharmacokinetics and comparable efficacy and safety compared to the intravenous formulations (Tan et al., 2021). The subcutaneous administration route is expected to reduce dosing complexity further and potentially enable home- or self-administration. Another example of a subcutaneous FDC is the coformulation of casirimab and imdevimab (REGEN-COV) that has just completed a phase 3 study demonstrating an 81% risk reduction of symptomatic SARS-CoV-2 infections of people that live in a household with SARS-CoV-2 infected individuals (REGENERON, 2021).

Other FDC programs are primarily early stage, with the majority of agents in preclinical development and a number already in phase 1 trials. The combinations being examined are mostly FDCs of novel agents, with programs spanning oncology, infectious disease, and ophthalmology therapy areas (Davey et al., 2016; Dienstmann et al., 2015; Džunková et al., 2016; Jacobsen et al., 2015; Kearns et al., 2015; Lieu et al., 2015; Lowy et al., 2010; Machiels et al., 2015; Nayak et al., 2014).

Dual-chamber devices

Developing an FDC from two or more biopharmaceuticals can be challenging due to potential difficulties in discriminating the active ingredients analytically and as the combined shelf-life, storage conditions, and stability are depending on the molecule with the strictest requirements.

A conceptional technology that might help overcoming these challenges for subcutaneous FDCs would be dual- or multiple-chamber devices. Such devices could ideally be loaded with different vials or cartridges containing the respective actives in separate formulations that would either be injected sequentially through the same needle or premixed and then injected together.

Such an approach would omit the need for long-term stability as required for an FDC and could enable shipment and storage in separate packages.

Subcutaneous dual-chamber devices on the market are typically filled with a powder and a diluent that are mixed directly prior to injection (LaRue & Malloy, 2015). Devices enabling the concomitant or sequential injection of two separate formulations are not yet available.

Timing of formulation and device lifecycle management

The most pertinent start of formulation and device lifecycle management activities in relation to a product's development stage is an important consideration for manufacturers. Pharmaceutical companies developing novel therapeutic entities naturally face budget and resource limitations leading to prioritized investments. During early development of a new biological drug focus is on demonstrating that the molecule is efficacious and safe in the target population. Aspects of optimizing the product presentation that aim at reducing dosing complexity are typically of lower priority at this point in time.

The decision on when to best optimize a given product with respect to its presentation and administration therefore becomes an important question. In particular, the following three scenarios need to be assessed: (1) introduce an optimized formulation or device directly at initial launch of the novel biopharmaceutical, (2) introduce product optimizations as a lifecycle management activity, or (3) no further optimization of the initial presentation will be required. The selection of the most viable scenario should be based on the planned treatment setting at launch and beyond, that is, with focus on the needs of the specific patient population and the overall healthcare system (Table 1.1).

Launch with optimized formulation and/or drug delivery device

Especially for disease-modifying agents with a novel mode of action, bringing the molecule to the patient is highest priority for a manufacturer and aspects related to drug delivery and the product presentation are typically of lower urgency. The situation can be different, if drug delivery plays a pivotal role in enabling drug administration. For example, performing intraocular injections can be a burden in ophthalmology departments (Bolme et al., 2020) and several types of drug formulation, such as gels, particles, and implants, are being developed to reduce dosing complexity and frequency to the extent

Table 1.1 Timing of formulation and device lifecycle management: the introduction of product optimizations in relation to initial launch of the biological drug is depending on whether drug delivery is required as an enabler for administration, on patient preferences and capabilities, as well as on the current and future treatment setting in the disease area.

Timing of product optimization	Rationale	Example
Introduce optimized formulation or device directly at initial launch of the novel biopharmaceutical	Drug delivery required as enabler for drug administration, that is, availability of reproducible dosing regimen mandated	Ocular and pulmonary drug delivery require convenient and ready-to-use device from initial launch onward
	Disease area where home- and self-administration represents standard of care	Subcutaneous self-administration with PFS or AI devices is broadly conducted in RA or MS
Introduce optimized formulation or device as a lifecycle management activity	Product optimizations could delay availability of novel potentially life-saving treatment	High-dose mAbs in oncology were initially launched with an intravenous dosing regimen; the subcutaneous dosing alternative did follow with the availability of appropriate high-volume technologies
	Changes in the treatment setting over time, such as the introduction of a flexible care setting	In MS, products were frequently introduced to the market with a vial or PFS presentation, followed by automated AI devices as lifecycle management
No further optimization of the initially launched product presentation required	Novel product is available with the most convenient formulation at initial molecule launch	Increasing understanding of the safety and tolerability of novel administration routes and the availability of advanced formulation and device platform technologies enables use of final drug delivery modality in pivotal trials

AI, autoinjector; *mAbs*, monoclonal antibodies; *MS*, multiple sclerosis; *PFS*, prefilled syringe; *RA*, rheumatoid arthritis.

feasible (Subrizi et al., 2019). Also, pulmonary administration mandates the availability of an inhalation device together with the product, and individual patients may require specific device types to achieve better compliance in daily treatment (Hristara-Papadopoulou et al., 2008). Therefore early availability of the most appropriate dosage forms allowing for a reproducible dosing regimen is critical.

Likewise, introducing a biological medicine into a disease area were silimar molecules with convenient drug administration regimens have already been established mandates a comparable drug presentation at initial launch. In such areas, manufacturers do their utmost to avoid introducing a greater dosing complexness for patients with prior experience with more convenient drug delivery forms. An example would be rheumatoid arthritis (RA), an area where subcutaneous administration with ready-to-use self-injection AI and PFP devices is standard-of-care since many years (van den Bemt et al., 2019). Here, a novel medication with a more invasive intravenous formulation but otherwise similar efficacy to available products would not be favored by patients and increase healthcare resources associated with drug administration.

Introduce optimized formulation and/or drug delivery device as lifecycle management

In the event that the timely launch of a new molecular entity is critical to provide patients with a novel potentially life-saving treatment alternative, optimizations in drug administration and presentation that could delay availability of the dug are typically conducted as a lifecycle management activity. In the recent past in oncology, for example, biopharmaceuticals were initially launched with an intravenous formulation. A subcutaneous dosing alternative was introduced sequentially at a later time point. Especially for high-dose biologics this step-wise approach did account for the development time of high-concentrated subcutaneous formulations and high-volume wearable injection devices (Bittner et al., 2018; Fettner et al., 2019).

Also, evolutions in the treatment setting in some disease areas, such as the introduction of a flexible care setting where patients and their nonprofessional caregivers can choose the place of drug administration (incl. home- or self-administration), typically mandate product optimizations that reduce dosing complexity. Biological injectables in RA or MS represent examples where ongoing optimizations in the administration route and drug delivery device are a way to improve patient convenience and compliance with the dosing instructions and thus to differentiate a product in a competitive market. Here, products are frequently introduced with a vial or PFS presentation, followed by a more user-friendly PFP or AI as lifecycle management soon after initial launch (Bittner & Schmidt, 2021).

No optimized formulation and/or drug delivery device required post initial launch

In the ideal case, a novel product is already available with the most convenient formulation at initial launch. With increasing understanding of the safety and tolerability of novel administration routes and the availability of advanced formulation and device platform technologies (for examples refer to paragraph "Technology platform approaches"), manufactures can leverage existing synergies between different product developments and apply them early on.

Technology platform approaches

Drug delivery technology platforms are defined as versatile applications suitable to a number of molecules or products. The importance and value of such platforms for the development of new biopharmaceutical medicines or for the lifecycle management of established treatments is growing rapidly.

Examples for established technology platforms

Examples for technology platforms are the Omnipod wearable OBDS or the Enhance formulation technology.

A wearable injection device technology platform is the Omnipod delivery system (Zisser, 2010). This patch device contains the drug in a PFS and the patch can be applied to multiple locations on the body (abdomen, arm, buttock, or leg), hence, facilitating administration in the home setting. Namely, the technology is applied for the subcutaneous administration of rapid-acting insulin. Insulin is delivered through an integrated soft cannula in the pod that is inserted automatically into the subcutaneous tissue. The patch is also used for pegfilgrastim (Neulasta), a granulocyte colony-stimulating factor therapy, that is given to patients the day after a chemotherapy cycle to boost the production of white blood cells that help fight infection (Burris et al., 2010). The single-use combination with the Omnipod was developed to allow patients to return home following their chemotherapy and thus to lessen the number of office visits while at the same time ensuring compliance with the treatment schedule (Yang et al., 2015).

The Enhance technology using recombinant human dispersion enhancer hyaluronidase to facilitate the subcutaneous bolus injection of larger volumes is an example for a platform supporting the development of subcutaneous dosing alternatives to established intravenous formulations while maintaining the dosing frequency of the intravenous route. In particular, high-dose biotherapeutics may benefit from this platform. While, previously, high-dose subcutaneous administration was typically done with a higher dosing frequency than the respective intravenous treatment to avoid the administration of high dosing volumes (Jolles et al., 2014), more recently, with this technology the frequency of administration can remain the same for the intravenous and subcutaneous versions of a biological medicine.

Leveraging platform synergies also for the underlying preclinical and clinical programs across different disease areas can reduce the development time for new molecules utilizing the same technology (Bittner et al., 2018).

Opportunities and barriers of applying and maintaining technology platforms

The benefits of using drug delivery technology platforms are manifold. Manufacturers can leverage synergies across their molecule portfolio resulting in faster and more cost and resource efficient development. Knowledge and data generated for one molecule in a specific disease area can largely be transferred to molecules in a different disease area using the same technology. This overall has the potential to increase the robustness of supporting data for registration, especially, after the approval of the first product that uses the technology platform.

In addition, there is a greater economy of scale by applying the same technology for a number of molecules. This includes more efficient use of manufacturing resources (equipment and labor) and can more easily justify investments for improvements or lifecycle management of the platform based on a larger number of products utilizing the same technology. It is anticipated that continuous advancement of a platform increases quality and reliability of the manufacturing processes and the respective products, which should result in higher customer satisfaction.

Technology platforms should also facilitate launch and commercialization activities. Marketing teams can leverage learnings from previously launched products and benefit from the increasing external familiarity with the platform. Patients and healthcare providers typically value the availability and continuous improvement of known technologies that have proven to deliver a product cost-efficiently and conveniently and fit with the daily routines and processes in healthcare institutions.

Conversely, a selected technology platform may not be the optimal technology solution for all pipeline molecules. There is also a risk that a technology platform may become a limiting factor for innovation depending on specific attributes, flexibility, or lack thereof. Another aspect to consider is the higher complexity when changing features of a technology platform, as typically all products using the same platform technology need to be assessed.

Opportunities and barriers inherent to the application of formulation and device platform technologies are summarized in Table 1.2.

Key aspects that define the value of drug delivery improvements

To ensure successful market adaptation, any drug delivery improvement needs to address obvious patient, healthcare provider, and societal needs (Table 1.3).

Patient needs

Today, with increasingly informed and well-educated patients there is a demand to more closely involve them as costumers into selecting the most appropriate dosing modality. This also includes the decision on where and how a drug is administered. Advances in drug delivery that remove complexities and inconvenience inherent to parenteral administration of biotherapeutics can play a pivotal role in achieving a truly patient-centric treatment management (Stroetmann, 2015). Namely, optimizations in dosing frequency and administration route, ready-to-use formulations and devices (Stegemann, 2018), as well as the application of connected or smart tools are important criteria in this context (Bittner et al., 2019). The treatment burden would be further reduced with an alignment of the dosing schedule for molecules administered as combination therapy. That is, manufacturers early on need to identify possible combination partners and develop new molecules with the same dosing frequency to the extent possible.

Table 1.2 Opportunities and barriers of using formulation and device platform technologies.

Opportunities	Barriers
Faster and more cost- and resource-efficient development by leveraging synergies across molecule portfolio	Selected technology platform may not be the optimal technology solution for all pipeline molecules resulting in a noncompetitive product
Increased robustness of data generated with platform	Risk that a technology platform may become a limiting factor for innovation depending on its technical features and degree of flexibility
Eased regulatory assessments, especially, after approval of the first product that uses the technology platform	Higher complexity when changing features of a technology platform
Greater economy of scale with more efficient use of manufacturing resources	
Increased robustness and quality of the manufacturing processes	
Facilitated launch and commercialization activities	

Independent of the underlying disease and treatment, patients have individual needs, capabilities, and preferences. While many patients value the opportunity to self-administer a medication at home, others would still prefer to be dosed by a healthcare provider in the hospital, office, or infusion center, to a large part due to the option for social interactions with other patients in the same area (McCuskey Shepley et al., 2012). Even the preference of one over another drug delivery device differs significantly. For example, in inhalation therapy, some patients prefer pMDIs over DPIs and the other way around (Dekhuijzen et al., 2016). For subcutaneous devices, the majority of patients prefer ready-to-use AI or PFP devices, but there is still a proportion of patients favoring PFSs (Vermeire et al., 2018). Consequently, to serve the various demands and to warrant compliance with the dosing regimen, more than one drug presentation or drug delivery route should be provided by manufacturers.

Table 1.3 Drug delivery improvements to address patient, healthcare provider, and societal needs.

Stakeholder	Need	How addressed
Patient	Involvement in selection of individual dosing modality and treatment setting	Reduced dosing complexity to increase patient experience (novel administration route, changes in dosing frequency, ready-to-use formulations and devices)
	Convenient drug administration regimen	Connected devices health apps to support home- and self-administration
	Choice of treatment setting	Availability of different presentations for the same drug to account for patient preferences and capabilities
Healthcare provider	Reassurance that novel drug presentations are efficacious and safe	Scientific data supporting that change in pharmacokinetic profile with new drug delivery modality does not impact established safety and efficacy profile
	Reassurance that patients adhere to the treatment regimen in the home setting and do not experience unexpected side effects	Connected devices and health apps to ensure compliance through adherence trackers, dosing reminders and data sharing tools
	Reimbursement for drug administration efforts	Application of value-based reimbursement models that reward efficiency, cost savings, and performance
Society	Reduced healthcare costs and resources	Simplified drug delivery to reduce treatment burden and improve patient experience
	Reduced loss of work time for patients and caregivers due to the time spent in a healthcare institution when receiving a medication	Adverse event tracking tools and electronic patient diaries to detect adverse events remotely and allow early treatment decisions
	Sustainable drug delivery	Reusable devices, less frequent dosing regimens and fixed doses to reduce wastage and environmental pollution

Healthcare provider needs

Healthcare providers to refer a novel drug delivery form, administration route, or dosing frequency to their patients require reinsurance that the changed product profile does not negatively impact the established safety and efficacy profile of the molecule. Key scientific evidence in this context can in many cases be based on the pharmacokinetic profile of the molecule. For example, clinical efficacious and tolerated dose levels of mAbs are typically selected to exceed a certain target concentration established from early animal models and clinical trials, as expressed by the minimal or trough concentration (C_{trough}) as a measure for predicting the receptor occupancy of the drug in the target tissue (Ahmed et al., 2018).

The dose levels for novel administration routes or changes in dosing frequency are subsequently calculated to result in noninferior C_{trough} or area under the serum concentration−time curve (AUC) compared to the initial drug presentation. At comparable C_{trough} or AUC, the pharmacokinetic parameter that could be affected by changing the drug delivery modality is the maximum plasma concentration (C_{max}). This is due to a different absorption rate into the systemic circulation with novel routes and drug delivery technologies or different individual dose levels with new dosing frequencies (Turner & Balu-Iyer, 2018). Consequently, available data from animal studies or clinical trials for the initial drug presentation need to provide evidence that C_{max} is not contributing to efficacy of the molecules. In the event that C_{max} is expected to impact efficacy, larger clinical trials are mandated to select an appropriate dose for the novel dosing regimen.

Particularly for parenteral administration, the actual time and resources spent for drug administration can pose a burden to healthcare providers and options to decrease dosing complexity are typically acknowledged. For example, a change from intravenous to subcutaneous dosing and home administration has been shown to markedly reduce the drug delivery-related healthcare institutional costs and resources. To feel comfortable with patients receiving their treatment at home, that is, outside of a controlled setting, healthcare providers still require real-world insights into whether dosing was accurately conducted and whether the medication was safe and well tolerated. Connected devices equipped with dosing reminders, adherence trackers, enabling a remote dialog between patients and the treating physician are expected to provide more transparency in this context. Applying such tools, the treating physician is informed in real time about the occurrence of adverse events. Patients can be called in for a clinic visit in time to treat episodes that may otherwise have resulted in hospitalization. Also, in the event of signs that a treatment is not efficacious, physicians would be enabled to make early decisions on changing treatments or to discuss the administration schedule with the patient. Long-term usage of connected tools still remains to be demonstrated on an individual patient and patient population basis (Bittner et al., 2019).

Key to successful adaptation of an improved product is that the responsible healthcare provider is appropriately reimbursed for drug administration-related efforts and that no additional burden is imposed. Thus the healthcare professional reimbursement model strongly impacts the value of formulation and device lifecycle management in a given legislation. While today, most reimbursement models are value-based and thus reward efficiency, cost savings, and performance, the so-called fee-for-service model rewards the volume and intensity of care and there are no incentives for healthcare providers to implement cost-saving measures (Adida et al., 2016).

Value-based reimbursement models (Henkel & Maryland, 2015), such as bundled payment (Agarwal et al., 2020), partial or full capitation (Goodson et al., 2001), pay-for-performance (Erdek, 2018), shared savings (Bailit & Hughes, 2011), or global budget (Huang et al., 2016), have incentives for lowering drug administration-related resources due to optimized drug delivery. The value for payers due to such product optimizations is the lowering of drug preparation and administration costs, reduction of drug wastage, avoidance of dosing errors, and related side effects that may result in rehospitalizations. Also, an overall improved patient experience with a more convenient formulation or drug delivery device is considered beneficial due to a positive impact on the compliance with the treatment regimen.

Fig. 1.2 compares the expected value of reduced drug administration complexity for the healthcare provider reimbursement models fee-for-service versus bundled payment as an example for a value-based reimbursement model.

Societal needs

Optimized drug delivery, thus improving the way a drug is administered to a patient, is expected to have a significant impact on a variety of societal

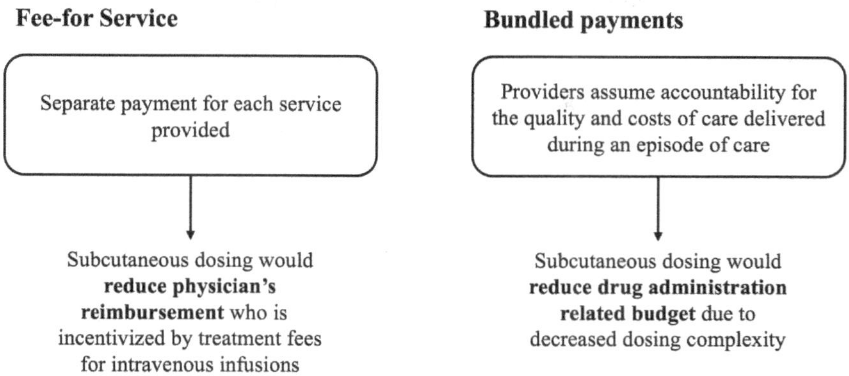

Figure 1.2 The reimbursement model drives the value of subcutaneous versus intravenous administration of biotherapeuticals. Expected value of reduced drug administration complexity for the healthcare provider reimbursement models fee-for-service versus bundled payment as an example for a value-based reimbursement model.

needs (Bittner et al., 2018). In particular, simplifications in drug administration that reduce the number of clinic visits, the related commute, and time spent at the unit are of notable societal value. With these improvements, patients and their nonprofessional caregivers lose less work and recreation time due to minimized travel to and time spent at a healthcare institution. Moreover, psychological barriers related to parenteral caregiver-assisted home- or self-administration can be reduced (Korytkowski, 2002).

To unfold the full societal benefit of improved drug delivery, manufacturers need to further invest into improving the education of healthcare providers, patients, and nonprofessional caregivers. Connected devices and health apps can be a valuable instrument within this framework. For the appropriate use of electronic tools to ensure adherence outside of a controlled setting, aspects, such as data protection laws and the individual patient's openness, to share data with a healthcare provider need to be taken into consideration.

While large parts of the patient and caregiver population are expected to welcome above improvements, there will not be a solution that fits the needs of all stakeholders. A viable focus of possible advancement in the context of sustainable drug delivery represents the fact that today the majority of drug delivery devices are for single-use only and thus contribute to wastage and pollution of the environment. Drug delivery devices that are, at least in parts, reusable are therefore a key focus area for device manufactures now and in the future (van den Bemt et al., 2019).

Summary and outlook

Formulation and device lifecycle management for biopharmaceuticals is conducted to reduce dosing complexity and improve convenience for molecules that typically need to be administered parenterally. Related optimizations provide more drug delivery options for patients and enable a flexible care setting, where patients and caregivers can choose the place of drug administration.

Prerequisites for successful adaptation of product optimizations are that drug delivery improvements address existing patient or healthcare needs, a proper reimbursement of the responsible healthcare provider, as well as access to the new product at reasonable costs. Consequently, the new improved product presentation is ideally available at price parity to the earlier version and there is evidence that usage reduces healthcare and societal costs and resources.

Key established improvements in the parenteral administration of biologics are the change from weight-adjusted intravenous to fixed dose subcutaneous dosing and the availability of FDCs and ready-to-use injection devices that facilitate dosing in a decentralized setting. Connected devices and health apps with dosing reminders, adherence trackers, offer the possibility for a remote dialog between patients and the treating physician and can further optimize home- and patient self-administration (Fig. 1.3).

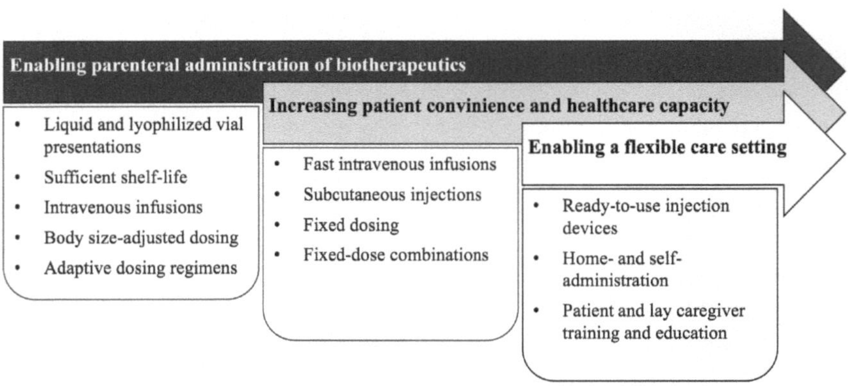

Figure 1.3 Evolution of parenteral drug administration for biopharmaceuticals. Drug delivery of high-dose biopharmaceuticals is changing from intravenous dosing regimens with predominantly body size-adjusted dosing regimens to subcutaneous injections with fixed doses and fixed-dose combinations. Today the development of ready-to-use injection (connected) devices supports drug administration in a flexible care setting.

Novel drug delivery approaches in pulmonary, brain, and ocular delivery are either already on the market or in late-stage clinical development and are a dosing alternative preferred by many patients compared to more invasive subcutaneous, intramuscular injections or intravenous infusions. Advances in the oral delivery of biotherapeutics are in development and, if proven to be feasible and tolerated, have to potential to introduce a real change in the treatment paradigm.

In the future, development approaches and technologies that complement combination therapy, such as FDCs or dual- and multiple-chamber devices, will be a key focus area of research to help improving clinical outcomes with biological combination therapies, while at the same time maintaining dosing convenience, costs, and sustainability.

References

Abramson, A., Caffarel-Salvador, E., Soares, V., Minahan, D., Tian, R. Y., Lu, X., Dellal, D., Gao, Y., Kim, S., Wainer, J., Collins, J., Tamang, S., Hayward, A., Yoshitake, T., Lee, H. C., Fujimoto, J., Fels, J., Frederiksen, M. R., Rahbek, U., & Traverso, G. (2019). A luminal unfolding microneedle injector for oral delivery of macromolecules. *Nature Medicine*, *25*, 1512–1518. Available from https://doi.org/10.1038/s41591-019-0598-9.

Adida, E., Mamani, H., & Nassiri, S. (2016). Bundled payment vs. fee-for-service: Impact of payment scheme on performance. *Management Science*, *63*(5), 1271–1656. Available from https://doi.org/10.1287/mnsc.2016.2445.

Afrezza. (2014). Highlights of prescribing information. https://www.afrezza.com/pdf/Full-Prescribing-Information.pdf.

Agarwal, R., Liao, L. M., Gupta, A., & Navathe, A. S. (2020). The impact of bundled payment on health care spending, utilization, and quality: A systematic review. *Health Affairs*, *39*(1), 50–57. Available from https://doi.org/10.1377/hlthaff.2019.00784.

Ahmed, S., Ellis, M., Li, H., Pallucchini, L., & Stein, A. M. (2018). Guiding dose selection of monoclonal antibodies using a new parameter (AFTIR) for characterizing ligand binding systems. *Journal of Pharmacokinetics and Pharmacodynamics*, *46*(3), 287–304. Available from https://doi.org/10.1007/s10928-019-09638-3.

Alexander, A., Agrawal, M., Uddin, A., Siddique, S., Shehata, A. M., Shaker, M. A., Rahman, S. A. U., Abdul, M. I. M., & Shaker, M. A. (2019). Recent expansions of novel strategies towards the drug targeting into the brain. *International Journal of Nanomedicine*, *14*, 5895–5909. Available from https://doi.org/10.2147/IJN.S210876.

Arden, M. A., Drabble, S., O'Cathain, A., Hutchings, M., & Wildman, M. (2019). Adherence to medication in adults with cystic fibrosis: An investigation using objective adherence data and the theoretical domains framework. *British Journal of Health Psychology*, *24*(2), 357–380. Available from https://doi.org/10.1111/bjhp.12357.

Ashley, G. W., Henise, J., Reid, R., & Santi, D. V. (2013). Hydrogel drug delivery system with predictable and tunable drug release and degradation rates. *Proceedings of the National Academy of Sciences of the United States of America*, *110*(6), 2318–2323. Available from https://doi.org/10.1073/pnas.1215498110.

Atkins, P. J. (2005). Dry powder inhalers: An overview. *Respiratory Care, 50*(10), 1304–1312. Available from https://pubmed.ncbi.nlm.nih.gov/16185366/.

Augustovski, F., Beratarrechea, A., Irazola, V., Rubinstein, F., Tesolin, P., Gonzalez, J., Lencina, V., Scolnik, M., Waimann, C., Navarta, D., Citera, G., & Soriano, E. R. (2013). Patient preferences for biologic agents in rheumatoid arthritis: A discrete-choice experiment. *Value Health, 16*(2), 385–393. Available from https://doi.org/10.1016/j.jval.2012.11.007.

Azevedo, V., Della, L., Troiano, C., Galli, N., Kleinfelder, A., Catolino, N. M., Urbano, P. C., & Martins, U. (2016). Adalimumab: A review of the reference product and biosimilars. *Biosimilars, 6*, 29–44. Available from https://doi.org/10.2147/BS.S98177.

Bailit, M., & Hughes, C. (2011). *Key design elements of shared-savings payment arrangements. The Commonwealth Fund* (20, pp. 1–6). Available from https://pubmed.ncbi.nlm.nih.gov/21847849/.

Bajracharya, R., Song, J. G., Back, S. Y., & Han, H. K. (2019). Recent advancements in non-invasive formulations for protein drug delivery. *Computional and Structural Biotechnology Journal, 17*, 1290–1308. Available from https://doi.org/10.1016/j.csbj.2019.09.004.

Bansal, D., Malla, S., Gudala, K., & Tiwari, P. (2013). Anti-counterfeit technologies: A pharmaceutical industry perspective. *Scientia Pharmaceutica, 81*(1), 1–13. Available from https://doi.org/10.3797/scipharm.1202-03.

Barclay, N., Tarallo, M., Hendrikx, T., & Marett, S. (2013). Patient preference for oral versus injectable and intravenous methods of treatment for rheumatoid arthritis. *Value in Health, 16*(7), PA568. Available from https://doi.org/10.1016/j.jval.2013.08.1521.

Barnett, A. H. (2004). Exubera inhaled insulin: A review. *The International Journal of Clinical Practice, 58*(4), 394–401. Available from https://doi.org/10.1111/j.1368-5031.2004.00178.x.

Barry, B. W. (2002). Drug delivery routes in skin: A novel approach. *Advances Drug Delivery Reviews, 54*(Suppl. 1), S31–S40. Available from https://doi.org/10.1016/s0169-409x(02)00113-8.

Bittner, B., Richter, W., & Schmidt, J. (2018). Subcutaneous administration of biotherapeutics: An overview of current challenges and opportunities. *BioDrugs, 32*(5), 425–440. Available from https://doi.org/10.1007/s40259-018-0295-0.

Bittner, B., & Schmidt, J. (2021). Subcutaneous drug delivery devices—Enablers of a flexible care setting. In E. Chappel (Ed.), *Drug delivery devices and therapeutic systems* (pp. 159–179). Elsevier. Available from https://doi.org/10.1016/B978-0-12-819838-4.00021-3.

Bittner, B., Schmit Chiesi, C., Kharawala, S., Kaur, G., & Schmidt, J. (2019). Connected drug delivery devices to complement drug treatments: Potential to facilitate disease management in home setting. *Medical Devices, 12*, 101–127. Available from https://doi.org/10.2147/MDER.S198943.

Boccara, F., Dent, R., Ruilope, L., & Valensi, P. (2017). Practical considerations for the use of subcutaneous treatment in the management of dyslipidaemia. *Advances in Therapy, 34*(8), 1876–1896. Available from https://doi.org/10.1007/s12325-017-0586-8.

Bolme, S., Sund Morken, T., Follestad, T., Lykke Sørensen, T., & Austeng, D. (2020). Task shifting of intraocular injections from physicians to nurses: A randomized single-masked noninferiority study. *Acta Ophthalmologica, 98*(2), 139–144. Available from https://doi.org/10.1111/aos.14184.

Borner, M. M., Schoffski, P., de Wit, R., Caponigro, F., Comella, G., Sulkes, A., Greim, G., Peters, G. J., van der Born, K., Wanders, J., de Boer, R. F., Martin, C., & Fumoleau, P. (2002). Patient preference and pharmacokinetics of oral modulated UFT versus intravenous fluorouracil and leucovorin: A randomised crossover trial in

advanced colorectal cancer. *European Journal of Cancer, 38*(3), 349–358. Available from https://doi.org/10.1016/s0959-8049(01)00371-9.
Burger, M., & Massoomi, F. (2019). IV insulin infusions: Is variation the enemy of quality and safety? *Pharmacy Practice News*. Available from https://www.pharmacypractice-news.com/Article/PrintArticle?articleID = 56740.
Burris, H., Belani, C., Kaufman, P., Gordon, A., Schwartzberg, L., Paroly, W., Shahi, S., Dreiling, L., & Saven, A. (2010). Pegfilgrastim on the same day versus next day of chemotherapy in patients with breast cancer, non-small-cell lung cancer, ovarian cancer, and non-Hodgkin's lymphoma: Results of four multicenter, double-blind, randomized phase II studies. *Journal of Oncology Practice, 6*(3), 133–140. Available from https://doi.org/10.1200/JOP.091094.
Bye, J. W., Platts, L., & Falconer, R. J. (2014). Biopharmaceutical liquid formulation: A review of the science of protein stability and solubility in aqueous environments. *Biotechnology Letters, 36*(5), 869–875. Available from https://doi.org/10.1007/s10529-013-1445-6.
Campochiaro, P. A., Marcus, D. M., Awh, C. C., Regillo, C., Adamis, A. P., Bantseev, V., Chiang, Y., Ehrlich, J. E., Erickson, S., Hanley, W. D., Horvath, J., Maass, K. F., Singh, N., Tang, F., & Barteselli, G. (2019). The port delivery system with ranibizumab for neovascular age-related macular degeneration: Results from the randomized phase 2 Ladder clinical trial. *Ophthalmology, 126*(8), 1141–1154. Available from https://doi.org/10.1016/j.ophtha.2019.03.036.
Can, M., Alibaz-Öner, F., Yılmaz-Öner, S., Atagündüz, P., İnanç, N., & Direskeneli, H. (2013). Accelerated infusion rates of rituximab are well tolerated and safe in rheumatology practice: A single-centre experience. *Clinical Rheumatology, 32*(1), 87–90. Available from https://doi.org/10.1007/s10067-012-2094-1.
Chan, A. H. Y., Harrison, J., Black, P. N., Mitchell, E. A., & Foster, J. M. (2015). Using electronic monitoring devices to measure inhaler adherence: A practical guide for clinicians. *The Journal of Allergy and Clinical Immunology: In Practice, 3*(3), 335–349. Available from https://doi.org/10.1016/j.jaip.2015.01.024.
Chang, D., Park, K., & Famili, A. (2019). Hydrogels for sustained delivery of biologics to the back of the eye. *Drug Discovery Today, 24*(8), 1470–1482. Available from https://doi.org/10.1016/j.drudis.2019.05.037.
Chung, C. H. (2008). Managing premedications and the risk for reactions to infusional monoclonal antibody therapy. *The Oncologist, 13*(6), 725–732. Available from https://doi.org/10.1634/theoncologist.2008-0012.
ClinicalTrials.gov. (2018a, September 19). A phase III study to evaluate the port delivery system with ranibizumab compared with monthly ranibizumab injections in participants with wet age-related macular degeneration (Archway). ClinicalTrials.gov. https://clinicaltrials.gov/ct2/show/NCT03677934?term = or + Delivery + System + With + Ranibizumab&draw = 2&rank = 3.
ClinicalTrials.gov. (2018b, September 4). A study of subcutaneous nivolumab monotherapy with or without recombinant human hyaluronidase PH20 (rHuPH20). ClinicalTrials.gov. https://clinicaltrials.gov/ct2/show/NCT03656718.
ClinicalTrials.gov. (2018c, September 11). Relative bioavailability study of subcutaneous injection versus intravenous infusion of pembrolizumab (MK-3475) in participants with advanced melanoma (MK-3475-555/KEYNOTE-555). ClinicalTrials.gov. https://clinicaltrials.gov/ct2/show/NCT03665597.
Copaxone. (2014, January 28). Highlights of prescribing information. FDA. https://www.accessdata.fda.gov/drugsatfda_docs/label/2014/020622s089lbl.pdf.
Cowan, R., Cohen, J. M., Rosenman, E., & Iyer, R. (2019). Physician and patient preferences for dosing options in migraine prevention. *The Journal of Headache and Pain, 20*, 50. Available from https://doi.org/10.1186/s10194-019-0998-8.

Crowe, T. P., Greenlee, M. H. W., Kanthasamy, A. G., & Hsu, W. H. (2018). Mechanism of intranasal drug delivery directly to the brain. *Life Sci.*, Feb 15; *195*, 44–52. Available from https://doi.org/10.1016/j.lfs.2017.12.025, Epub 2017 Dec 22. PMID: 29277310.

Da Rocha, S. R. P., Bharatwaj, B., & Saiprasad, S. (2011). Science and technology of pressurized metered-dose inhalers. In H. D. C. Smyth, & A. J. Hickey (Eds.), *Controlled pulmonary drug delivery. Advances in delivery science and technology*. Springer. Available from https://doi.org/10.1007/978-1-4419-9745-6_8.

Davey, R. T., Dodd, L., Proschan, M. A., Neaton, J., Neuhaus Nordwall, J., Koopmeiners, J. S., Beigel, J., Tierney, J., Lane, H. C., Fauci, A. S., Massaquoi, M. B. F., Sahr, F., & Malvy, D. (2016). A randomized, controlled trial of ZMapp for ebola virus infection. *The New England Journal of Medicine*, *375*(15), 181–193. Available from https://doi.org/10.1056/NEJMoa1604330.

Dekhuijzen, P. N. R., Lavorini, F., & Usmani, O. S. (2016). Patients' perspectives and preferences in the choice of inhalers: The case for Respimat® or HandiHaler®. *Patient Preference and Adherence*, *10*, 1561–1572. Available from https://doi.org/10.2147/PPA.S82857.

DeVos, S. L., & Miller, T. M. (2013). Direct intraventricular delivery of drugs to the rodent central nervous system. *Journal of Visualized Experiments*, *75*, e50326. Available from https://doi.org/10.3791/50326.

Dibonaventura, M. D., Wagner, J. S., Girman, C. J., Brodovicz, K., Zhang, Q., Qiu, Y., Pentakota, S. R., & Radican, L. (2010). Multinational Internetbased survey of patient preference for newer oral or injectable type 2 diabetes medication. *Patient Preference and Adherence*, *4*, 397–406. Available from https://doi.org/10.2147/PPA.S14477.

Dienstmann, R., Patnaik, A., Garcia-Carbonero, R., Cervantes, A., Benavent, M., Roselló, S., Tops, B. B., van der Post, R. S., Argilés, G., Skartved, N. J., Hansen, U. H., Hald, R., Pedersen, M. W., Kragh, M., Horak, I. D., Braun, S., Van Cutsem, E., Tolcher, A. W., & Tabernero, J. (2015). Safety and activity of the first-in-class Sym004 anti-EGFR antibody mixture in patients with refractory colorectal cancer. *Cancer Discovery*, *5*(6), 598–609. Available from https://doi.org/10.1158/2159-8290.CD-14-1432.

Doessegger, L., & Banholzer, M. L. (2015). Clinical development methodology for infusion-related reactions with monoclonal antibodies. *Clinical and Translational Immunology*, *4*(7), e39. Available from https://doi.org/10.1056/NEJMoa1607751.

Džunková, M., D'Auria, G., Xu, H., Huang, J., Duan, Y., Moya, A., Kelly, C. P., & Chen, X. (2016). The monoclonal antitoxin antibodies (actoxumab-bezlotoxumab) treatment facilitates normalization of the gut microbiota of mice with clostridium difficile infection. *Frontiers in Cellular and Infection Microbiology*, *6*, 119. Available from https://doi.org/10.3389/fcimb.2016.00119.

Eagle Pharmaceuticals. (2016). Eagle Pharmaceuticals signs definitive agreement to acquire Arsia Therapeutics marking entry into biosimilar market. https://www.businesswire.com/news/home/20161111005045/en/Eagle-Pharmaceuticals-Signs-Definitive-Agreement-Acquire-Arsia.

Edelman, B. (2004). Specialty pharmacy booms, but for how long? *Biotechnology Healthcare*, *1*(1), 31–37. Available from https://www.ncbi.nlm.nih.gov/pmc/articles/PMC3555159/.

Egorin, M. J. (2003). Horseshoes, hand grenades, and body-surface area-based dosing: Aiming for a target. *Journal of Clinical Oncology*, *21*(2), 182–183. Available from https://doi.org/10.1200/JCO.2003.10.084.

Erdek, M. A. (2018). Pay-for-performance reimbursement for clinicians: Common sense or a wolf in sheep's clothing? *Pain Medicine*, *19*(11), 2106–2108. Available from https://doi.org/10.1093/pm/pny168.

Erdő, F., Borsa, L. A., Farkasa, D., Bajzaa, A., & Gizurarsonb, S. (2018). Evaluation of intranasal delivery route of drug administration for brain targeting. *Brain Research Bulletin*, *143*, 155–170. Available from https://doi.org/10.1016/j.brainresbull.2018.10.009.

Eylea. (2019). Highlights of prescribing information. https://www.accessdata.fda.gov/drugsatfda_docs/label/2019/125387s061lbl.pdf.

Fasano, A. (1998). Innovative strategies for the oral delivery of drugs and peptides. *Trends in Biotechnology*, *16*(4), 152–157. Available from https://doi.org/10.1016/S0167-7799(97)01170-0.

Felip, E., Burotto, M., Zvirbule, Z., Herraez-Baranda, L. A., Chanu, P., Kshirsagar, S., Maiya, V., Chang, P., Pozzi, E., Marchand, M., Monchalin, M., Tanaka, K., Tosti, N., Wang, B., & Restuccia, E. (2021). Results of a dose-finding phase 1b study of subcutaneous atezolizumab in patients with locally advanced or metastatic non–small cell lung cancer. *Clinical Pharmacology in Drug Develop*. Available from https://doi.org/10.1002/cpdd.936.

Fettner, S., Mela, C., Wildenhahn, F., Tavanti, M., Wells, C., Douglass, W., & Mallalieu, N. L. (2019). Evidence of bioequivalence and positive patient user handling of a tocilizumab autoinjector. *Expert Opinion on Drug Delivery*, *16*(5), 551–561. Available from https://doi.org/10.1080/17425247.2019.1604678.

Fjellestad-Paulsen, A., Wille, S., & Harris, A. S. (1987). Comparison of intranasal and oral desmopressin for nocturnal enuresis. *Archives of Disease in Childhood*, *62*(7). Available from https://doi.org/10.1136/adc.62.7.674, 674–667.

FluMist Quadrivalent. (2019). Highlights of prescribing information. https://www.fda.gov/media/83072/download.

Fluzone Quadrivalent. (2020). Highlights of prescribing information. https://www.fda.gov/media/119856/download.

Freshwater, T., Kondic, A., Ahamadi, M., Li, C. H., de Greef, R., de Alwis, D., & Stone, J. A. (2017). Evaluation of dosing strategy for pembrolizumab for oncology indication. *Journal for Immunotherapy of Cancer*, *5*(43). Available from https://doi.org/10.1186/s40425-017-0242-5.

Frost, G. (2007). Recombinant human hyaluronidase (rHuPH20): An enabling platform for subcutaneous drug and fluid administration. *Expert Opinion on Drug Delivery*, *4*(4), 427–440. Available from https://doi.org/10.1517/17425247.4.4.427.

Garidel, P., Kuhn, A. B., Schäfer, L. V., Karow-Zwick, A. R., & Blech, M. (2017). High-concentration protein formulations: How high is high? *European Journal of Pharmaceutics and Biopharmaceutics*, *119*, 353–360. Available from https://doi.org/10.1016/j.ejpb.2017.06.029.

Gaudana, R., Ananthula, H. K., Parenky, A., & Mitra, A. K. (2010). Ocular drug delivery. *Journal of the American Association of Pharmaceutical Scientists*, *12*(3), 348–360. Available from https://doi.org/10.1208/s12248-010-9183-3.

Gervasi, V., Dall Agnol, R., Cullen, S., McCoy, T., Vucen, S., & Crean, A. (2018). Parenteral protein formulations: An overview of approved products within the european union. *European Journal of Pharmaceutics and Biopharmaceutics*, *131*, 8–24. Available from https://doi.org/10.1016/j.ejpb.2018.07.011.

Gildon, B. W. (2018). InPen smart insulin pen system: Product review and user experience. *Diabetes Spectrum*, *31*(4), 354–358. Available from https://doi.org/10.2337/ds18-0011.

Goodson, J. D., Bierman, A. S., Fein, O., Rask, K., Rich, E. C., & Selker, H. P. (2001). The future of capitation: The physician role in managing change in practice. *Journal of General Internal Medicine*, *16*(4), 250–256. Available from https://doi.org/10.1046/j.1525-1497.2001.016004250.x.

Graves, S., Lei, L., Melamed, B., Pinedo, M., Qi, L., Shen, Z. J., & Xu, X. (2009). New challenges to emergency management of pharmaceutical/healthcare supply chain disruptions. DHS Workshop on Incident Management, Resource Management, and

Supply Chain Management. https://cert.ics.uci.edu/EMWS09/presentations/Position %20Papers/emws09_submission_2.pdf
Grzybowski, A., Told, R., Sacu, S., Bandello, F., Moisseiev, E., Loewenstein, A., & Schmidt-Erfurth, U. (2018). 2018 update on intravitreal injections: Euretina expert consensus recommendations. *Ophthalmologica*, *239*(4), 181–193. Available from https://doi.org/10.1159/000486145.
Gupta, S., Jain, A., Chakraborty, M., Sahni, J. K., Ali, J., & Dang, S. (2013). Oral delivery of therapeutic proteins and peptides: A review on recent developments. *Drug Delivery*, *20*(6), 237–246. Available from https://doi.org/10.3109/10717544.2013.819611.
Haeuser, C., Goldbach, P., Huwyler, J., Friess, W., & Allmendinger, A. (2020). Impact of dextran on thermal properties, product quality attributes, and monoclonal antibody stability in freeze-dried formulations. *European Journal of Pharmaceutics and Biopharmaceutics*, *147*, 45–56. Available from https://doi.org/10.1016/j.ejpb.2019.12.010.
Hamman, J. H., Enslin, G. M., & Kotzé, A. F. (2005). Oral delivery of peptide drugs: Barriers and developments. *BioDrugs*, *19*(3), 165–177. Available from https://doi.org/10.2165/00063030-200519030-00003.
Hashim, H., & Abrams, P. (2008). Desmopressin for the treatment of adult nocturia. *Therapy*, *5*(5), 667–683. Available from https://www.openaccessjournals.com/articles/desmopressin-for-the-treatment-of-adult-nocturia.pdf.
Hassan, M., Bonafede, M. M., Limone, B., Hodgkins, P., & Sawicki, G. (2016). The burden of cystic fibrosis: Pulmonary exacerbations and health care resource utilization in a commercially insured population in the United States. *Value in Health*, *19*(3), A115. Available from https://doi.org/10.1016/j.jval.2016.03.452.
Heinemann, L. (2008). The failure of Exubera: Are we beating a dead horse? *Journal of Diabetes Science and Technology*, *2*(3), 518–529. Available from https://doi.org/10.1177/193229680800200325.
Hendrikx, J. J. M. A., Haanen, J. B. A. G., Voest, E. E., Schellens, J. H. M., Huitema, A. D. R., & Beijnen, J. H. (2017). Fixed dosing of monoclonal antibodies in oncology. *The Oncologist*, *22*(10), 1212–1221. Available from https://doi.org/10.1634/theoncologist.2017-0167.
Henkel, R. J., & Maryland, P. A. (2015). The risks and rewards of value-based reimbursement. *Frontiers of Health Services Management*, *32*(2), 3–16. Available from https://pubmed.ncbi.nlm.nih.gov/26817266/.
Henricks, L. M., Schellens, J. H., Huitema, A. D., & Beijnen, J. H. (2015). The use of combinations of monoclonal antibodies in clinical oncology. *Cancer Treatment Reviews*, *41*(10), 859–867. Available from https://doi.org/10.1016/j.ctrv.2015.10.008.
Hinderer, C., Katz, N., Dyer, C., Goode, T., Johansson, J., Bell, P., Richman, L., Buza, E., & Wilson, J. M. (2020). Translational feasibility of lumbar puncture for intrathecal AAV administration. *Molecular Therapy: Methods and Clinical Development*, *17*, 969–974. Available from https://doi.org/10.1016/j.omtm.2020.04.012.
Holz, F. G., Tadayoni, R., Beatty, S., Berger, A., Cereda, M. G., Cortez, R., Hoyng, C. B., Hykin, P., Staurenghi, G., Heldner, S., Bogumil, T., Heah, T., & Sivaprasad, S. (2015). Multi-country real-life experience of anti-vascular endothelial growth factor therapy for wet age-related macular degeneration. *The British Journal of Ophthalmology*, *99*(2), 220–226. Available from https://doi.org/10.1136/bjophthalmol-2014-305327.
Hristara-Papadopoulou, A., Tsanakas, J., Diomou, G., & Papadopoulou, O. (2008). Current devices of respiratory physiotherapy. *Hippokratia*, *12*(4), 211–220. Available from https://www.ncbi.nlm.nih.gov/pmc/articles/PMC2580042/.
Huang, Y., Liu, Y., Yang, X., Li, J., & Fang, P. (2016). Global budget payment system helps to reduce outpatient medical expenditure of hypertension in china. *SpringerPlus*, *5*(1), 1877. Available from https://doi.org/10.1186/s40064-016-3565-7.

Hyqvia. (2014). Highlights of prescribing information. https://www.fda.gov/media/89844/download.

Iyer, S., Radwan, A. E., Hafezi-Moghadam, A., Malyala, P., & Amiji, M. (2019). Long-acting intraocular delivery strategies for biological therapy of age-related macular degeneration. *Journal of Controlled Release, 296*, 140–149. Available from https://doi.org/10.1016/j.jconrel.2019.01.007.

Jacobsen, H. J., Poulsen, T. T., Dahlman, A., Kjær, I., Koefoed, K., Sen, J. W., Weilguny, D., Bjerregaard, B., Andersen, C. R., Horak, I. D., Pedersen, M. W., Kragh, M., & Lantto, J. (2015). Pan-HER, an antibody mixture simultaneously targeting EGFR, HER2, and HER3, effectively overcomes tumor heterogeneity and plasticity. *Clinical Cancer Research, 21*(18), 4110–4122. Available from https://doi.org/10.1158/1078-0432.CCR-14-3312.

Jain, S., Patel, N., Shah, M. K., Khatri, P., & Vora, N. (2017). Recent advances in lipid-based vesicles and particulate carriers for topical and transdermal application. *Journal of Pharmaceutical Sciences, 106*(2), 423–445. Available from https://doi.org/10.1016/j.xphs.2016.10.001.

Jena, L., McErlean, E., & McCarthy, H. (2020). Delivery across the blood-brain barrier: Nanomedicine for glioblastoma multiforme. *Drug Delivery and Translational Research, 10*(2), 304–318. Available from https://doi.org/10.1007/s13346-019-00679-2.

Jolles, S., Orange, J. S., Gardulf, A., Stein, M. R., Shapiro, R., Borte, M., & Berger, M. (2014). Current treatment options with immunoglobulin G for the individualization of care in patients with primary immunodeficiency disease. *Clinical and Experimental Immunology, 179*(2), 146–160. Available from https://doi.org/10.1111/cei.12485.

Kalra, S., Gupta, N., Bhatia, V., Amin, F., & Unnikrishnan, A. G. (2019). Patient-centered choice of prandial insulin. *United States Endocrinology, 15*(1), 20–23. Available from https://doi.org/10.17925/USE.2019.15.1.20.

Kang, S. P., & Saif, M. W. (2007). Infusion-related and hypersensitivity reactions of monoclonal antibodies used to treat colorectal cancer—Identification, prevention, and management. *The Journal of Supportive Oncology, 5*, 451–457. Available from https://pubmed.ncbi.nlm.nih.gov/26547132/.

Katz, N., Goode, T., Hinderer, C., Hordeaux, J., & Wilson, J. M. (2018). Standardized method for intra-cisterna magna delivery under fluoroscopic guidance in nonhuman primates. *Human Gene Therapy Methods, 29*(5), 212–219. Available from https://doi.org/10.1089/hgtb.2018.041.

Kearns, J. D., Bukhalid, R., Sevecka, M., Tan, G., Gerami-Moayed, N., Werner, S. L., Kohli, N., Burenkova, O., Sloss, C. M., King, A. M., Fitzgerald, J. B., Nielsen, U. B., & Wolf, B. B. (2015). Enhanced targeting of the EGFR network with MM-151, an oligoclonal anti-EGFR antibody therapeutic. *Molecular Cancer Therapeutics, 14*(7), 1625–1636. Available from https://doi.org/10.1158/1535-7163.MCT-14-0772.

Kim, S. J., Kim, K. W., Shin, Y. K., Kwon, J. W., Kang, H. Y., Park, Y. A., Shin, J. Y., Kim, S. Y., & Han, W. Y. (2019). In use stability of the rituximab biosimilar CT-P10 (truxima®) following preparation for intravenous infusion and storage. *BioDrugs, 33*(2), 221–228. Available from https://doi.org/10.1007/s40259-019-00336-7.

Klonoff, D. C., & Kerr, D. (2018). Smart pens will improve insulin therapy. *Journal of Diabetes Science and Technology, 12*(3), 551–553. Available from https://doi.org/10.1177/1932296818759845.

Korytkowski, M. (2002). When oral agents fail: Practical barriers to starting insulin. *International Journal of Obesity, 26*(Suppl 3), S18–S24. Available from https://doi.org/10.1038/sj.ijo.0802173.

Larson, A.M., Love, E., Weight, A.K., Crane, A., Langer, R.S., & Alexander, M. (2015). Liquid protein formulations containing viscosity-lowering agents. United States patent

2015/0071925 A1. https://patentimages.storage.googleapis.com/16/59/a0/4c3221c4c0f848/UnitedStates20150071925A1.pdf.

LaRue, S., & Malloy, J. (2015). Evaluation of the dual-chamber pen design for the injection of exenatide once weekly for the treatment of type 2 diabetes. *Journal of Diabetes Science and Technology*, *9*(4), 815–821. Available from https://doi.org/10.1177/1932296815576186.

Lau, P. E. T., Jenkins, K. S., & Layton, C. J. (2018). Current evidence for the prevention of endophthalmitis in anti-VEGF intravitreal injections. *Hindawi Journal of Ophthalmology*. Available from https://doi.org/10.1155/2018/8567912.

Lenz, H. J. (2007). Management and preparedness for infusion and hypersensitivity reactions. *The Oncologist*, *12*(5), 601–609. Available from https://doi.org/10.1634/theoncologist.12-5-601.

Li, J., & Mooney, D. J. (2016). Designing hydrogels for controlled drug delivery. *Nature Reviews Materials*. Available from https://doi.org/10.1038/natrevmats.2016.71.

Lieu, C. H., Beeram, M., Harb, W. A., Kearns, J. D., Sloss, C. M., Nering, R., Adjei, A. A., & Wolf, B. B. (2015). Safety, pharmacology, and preliminary clinical activity of MM-151: An oligocolnal anti-EGFR theraputic in patients with cetuximab-resistant CRC and other refractory solid tumors. *Journal of Clinical Oncology*, *33*(3 Suppl), A647. Available from https://ascopubs.org/doi/abs/10.1200/jco.2015.33.3_suppl.647.

Lowy, I., Molrine, D. C., Leav, B. A., Blair, B. M., Baxter, R., Gerding, D. N., Nichol, G., Thomas, W. D., Leney, M., Sloan, S., Hay, C. A., & Ambrosino, D. M. (2010). Treatment with monoclonal antibodies against clostridium difficile toxins. *The New England Journal of Medicine*, *362*(3), 197–205. Available from https://doi.org/10.1056/NEJMoa0907635.

Lucentis. (2017). Highlights of prescribing information. https://www.gene.com/download/pdf/lucentis_prescribing.pdf.

Lugaresi, A. (2013). RebiSmartTM (version 1.5) device for multiple sclerosis treatment delivery and adherence. *Expert Opinion on Drug Delivery*, *10*(2), 273–283. Available from https://doi.org/10.1517/17425247.2013.746311.

Machiels, J. P., Specenier, P., Krauß, J., Dietz, A., Kaminsky, M. C., Lalami, Y., Henke, M., Keilholz, U., Knecht, R., Skartved, N. J., Horak, I. D., Pamperin, P., Braun, S., & Gauler, T. C. (2015). A proof of concept trial of the anti-EGFR antibody mixture Sym004 in patients with squamous cell carcinoma of the head and neck. *Cancer Chemotherapy and Pharmacology*, *76*(1), 13–20. Available from https://doi.org/10.1007/s00280-015-2761-4.

Macugen. (2011). Highlights of prescribing information. https://www.accessdata.fda.gov/drugsatfda_docs/label/2011/021756s018lbl.pdf.

McCuskey Shepley, M., Rybkowski, Z., Aliber, J., & Lange, C. (2012). Ambulatory infusion suite: Pre- and post-occupancy evaluation. *Building Research and Information*, *40*(6), 700–712. Available from https://doi.org/10.1080/09613218.2012.709372.

McLenon, J., & Rogers, M. A. M. (2019). The fear of needles: A systematic review and meta-analysis. *Journal of Advanced Nursing*, *75*(1), 30–42. Available from https://doi.org/10.1111/jan.13818.

Mecklenburg, L., & Schraermeyer, U. (2007). An overview on the toxic morphological changes in the retinal pigment epithelium after systemic compound administration. *Toxicologic Pathology*, *35*(2), 252–267. Available from https://doi.org/10.1080/01926230601178199.

Melmed, S., Popovic, V., Bidlingmaier, M., Mercado, M., van der Lely, A. J., Biermasz, N., Bolanowski, M., Coculescu, M., Schopohl, J., Racz, K., Glaser, B., Greenman, Y., Trainer, P., Mezosi, E., Shimon, I., Giustina, A., Korbonits, M., Bronstein, M. D., Kleinberg, D., & Strasburger, C. (2015). Safety and efficacy of oral octreotide in acromegaly: Results of a multicenter phase III trial. *The Journal of Clinical Endocrinology and Metabolism, 100*(4), 1699–1708. Available from https://doi.org/10.1210/jc.2014-4113.

Menditto, E., Orlando, V., De Rosa, G., Minghetti, P., Musazzi, U. M., Cahir, C., Kurczewska-Michalak, M., Kardas, P., Costa, E., & Manuel, J. (2020). Patient centric pharmaceutical drug product design—The impact on medication adherence. *Pharmaceutics, 12*(1), 44. Available from https://doi.org/10.3390/pharmaceutics12010044.

Merchant, R., Inamdar, R., Henderson, K., Barrett, M., Su, J. G., Riley, J., Van Sickle, D., & Stempel, D. (2018). Digital health intervention for asthma: Patient-reported value and usability. *Journal of Medical Internet Research MHealth UHealth, 6*(6), e133. Available from https://doi.org/10.2196/mhealth.7362.

Microglassification. (2019). https://www.lindybio.com/technology.

Mitri, J., & Pittas, A. G. (2009). Inhaled insulin—What went wrong. *Nature Clinical Practice Endocrinology and Metabolism, 5*(1), 24–25. Available from https://doi.org/10.1038/ncpendmet1007.

Monés, J., Singh, R. P., Bandello, F., Souied, E., Liu, X., & Gale, R. (2020). Undertreatment of neovascular age-related macular degeneration after 10 years of anti-vascular endothelial growth factor therapy in the real world: The need for a change of mindset. *Ophthalmologica, 243*(1), 1–8. Available from https://doi.org/10.1159/000502747.

Morales, J. O., Fathe, K. R., Brunaugh, A., Ferrati, S., Li, S., Montenegro-Nicolini, M., Mousavikhamene, Z., McConville, J. T., Prausnitz, M. R., & Smyth, H. D. C. (2017). Challenges and future prospects for the delivery of biologics: Oral mucosal, pulmonary, and transdermal routes. *Journal of the American Association of Pharmaceutical Scientists, 19*(3), 652–668. Available from https://doi.org/10.1208/s12248-017-0054-z.

Naehrig, S., Lang, S., Schiffl, H., Huber, R. M., & Fischer, R. (2011). Lung function in adult patients with cystic fibrosis after using the eFlow® rapid for one year. *European Journal of Medical Research, 16*(2), 63–66. Available from https://doi.org/10.1186/2047-783X-16-2-63.

Nayak, S. U., Griffiss, J. M., McKenzie, R., Fuchs, E. J., Jurao, R. A., An, A. T., Ahene, A., Tomic, M., Hendrix, C. W., & Zenilman, J. M. (2014). Safety and pharmacokinetics of XOMA 3AB, a novel mixture of three monoclonal antibodies against botulinum toxin A. *Antimicrobial Agents and Chemotherapy, 58*(9), 5047–5053. Available from https://doi.org/10.1128/AAC.02830-14.

Newswanger, B., Ammons, S., Phadnis, N., Ward, K., Castle, J., Campbell, R. W., & Prestrelski, S. J. (2015). Development of a highly stable, nonaqueous glucagon formulation for delivery via infusion pump systems. *Journal of Diabetes Science and Technology, 9*(1), 24–33. Available from https://doi.org/10.1177/1932296814565131.

Norman, P. (2016). Humira: The impending patent battles over adalimumab biosimilars. *Pharmaceutical Patent Analyst, 5*(3). Available from https://doi.org/10.4155/ppa-2016-0002.

Oleck, J., Kassam, S., & Goldman, J. D. (2016). Commentary: Why was inhaled insulin a failure in the market? *Diabetes Spectrum, 29*(3), 180–184. Available from https://doi.org/10.2337/diaspect.29.3.180.

ONdrugDelivery. (2019). Enabling subcutaneous delivery of biologics. https://www.ondrugdelivery.com/wp-content/uploads/2019/05/97_May_2019_Elektrofi.pdf.

Ovacik, M., & Lin, K. (2018). Tutorial on monoclonal antibody pharmacokinetics and its considerations in early development. *Journal of Clinical and Translational Science, 11*(6), 540–552. Available from https://doi.org/10.1111/cts.12567.

Ozempic. (2017). Highlights of prescribing information. https://www.accessdata.fda.gov/drugsatfda_docs/label/2017/209637lbl.pdf.

Pardridge, W. M. (2020). Blood-brain barrier and delivery of protein and gene therapeutics to brain. *Frontiers in Aging Neuroscience*, *11*, 373. Available from https://doi.org/10.3389/fnagi.2019.00373.

Park, K., Kwon, I. C., & Park, K. (2011). Oral protein delivery: Current status and future prospect. *Reactive and Functional Polymers*, *71*(3), 280–287. Available from https://doi.org/10.1016/j.reactfunctpolym.2010.10.002.

Pascolini, D., & Mariotti, S. P. (2012). Global estimates of visual impairment: 2010. *British Journal of Ophthalmology*, *96*(5), 614–618. Available from https://doi.org/10.1136/bjophthalmol-2011-300539.

Patel, A. J., Pieramici, D. J., & Bagheri, N. (2021). Evaluation of port delivery system with ranibizumab for the treatment of neovascular age-related macular degeneration. *Therapeutic Delivery*, *12*(3), 191–200. Available from https://doi.org/10.4155/tde-2020-0132.

Patil, J. S., & Sarasija, S. (2012). Pulmonary drug delivery strategies: A concise, systematic review. *Lung India*, *29*(1), 44–49. Available from https://doi.org/10.4103/0970-2113.92361.

Patton, J. S., & Byron, P. R. (2007). Inhaling medicines: Delivering drugs to the body through the lungs. *Nature Reviews Drug Discovery*, *6*(1), 67–74. Available from https://doi.org/10.1038/nrd2153.

Phesgo. (2020). Highlights of prescribing information. https://www.accessdata.fda.gov/drugsatfda_docs/label/2020/761170s000lbl.pdf.

Philippart, M., Schmidt, J., & Bittner, B. (2016). Oral delivery of therapeutic proteins and peptides: An overview of current technologies and recommendations for bridging from approved intravenous or subcutaneous administration to novel oral regimens. *Drug Research*, *66*(3), 113–120. Available from https://doi.org/10.1055/s-0035-1559654.

Pulmozyme. (2018). Highlights of prescribing information. https://www.gene.com/download/pdf/pulmozyme_prescribing.pdf.

Quartino, A. L., Li, H., Jin, J. Y., Wada, D. R., Benyunes, M. C., McNally, V., Viganò, L., Nijem, I., Lum, B. L., & Garg, A. (2017). Pharmacokinetic and exposure-response analyses of pertuzumab in combination with trastuzumab and docetaxel during neoadjuvant treatment of HER2+ early breast cancer. *Cancer Chemotherapy and Pharmacology*, *79*(2), 353–361. Available from https://doi.org/10.1007/s00280-016-3218-0.

Rees, F., Packham, A., Blake, C., Hills, E., Scutt, G., & St Clair-Jones, A. (2020). Can we reduce intravenous monoclonal antibody observation times without compromising patient safety? A single centre retrospective observational study. *Journal of Crohn's and Colitis, 14, Suppl. 1*, S504–S505. Available from https://doi.org/10.1093/ecco-jcc/jjz203.732.

REGENERON. (2021). Phase 3 prevention trial with subcutaneous administration of REGEN-COV. https://investor.regeneron.com/news-releases/news-release-details/phase-3-prevention-trial-showed-81-reduced-risk-symptomatic-sars.

Reinisch, W., Jang, B. I., Borzan, V., Lahat, A., Osipenko, M., Mostovoy, Y., Schreiber, S., Ben-Horin, S., Lee, S. J., Suh, J. H., Lee, S. G., & Ye, B. D. (2019). A novel formulation of CT-P13 (infliximab biosimilar) for subcutaneous administration: 1-Year result from a phase I open-label randomised controlled trial in patients with active Crohn's disease. *Journal of Crohn's and Colitis*, *13*(1), S066–S077. Available from https://doi.org/10.1093/ecco-jcc/jjy222.096.

Renukuntla, J., Vadlapudi, A. D., Patel, A., Boddu, S. H., & Mitra, A. K. (2013). Approaches for enhancing oral bioavailability of peptides and proteins. *International Journal of Pharmaceutics*, *447*(1–2), 75–93. Available from https://doi.org/10.1016/j.ijpharm.2013.02.030.

Richter, W. F., & Jacobsen, B. (2014). Subcutaneous absorption of biotherapeutics: Knowns and unknowns. *Drug Metabolism and Disposition*, *42*(11), 1881–1889. Available from https://doi.org/10.1124/dmd.114.059238.

Rodrigues, G. A., Lutz, D., Shen, J., Yuan, X., Shen, H., Cunningham, J., & Rivers, H. M. (2018). Topical drug delivery to the posterior segment of the eye: Addressing the challenge of preclinical to clinical translation. *Pharmaceutical Research, 35*(12), 245. Available from https://doi.org/10.1007/s11095-018-2519-x.

Rohrer, J., Lupo, N., & Bernkop-Schnürch, A. (2018). Advanced formulations for intranasal delivery of biologics. *International Journal of Pharmaceutics, 553*(1−2), 8−20. Available from https://doi.org/10.1016/j.ijpharm.2018.10.029.

Rombouts, M. D., Swart, E. L., Van Den Eertwegh, A. J. M., & Crul, M. (2020). Systematic review on infusion reactions to and infusion rate of monoclonal antibodies used in cancer treatment. *Anticancer Research, 40*(3), 1201−1218. Available from https://doi.org/10.21873/anticanres.14062.

Rybelsus. (2017). Highlights of prescribing information. https://www.accessdata.fda.gov/drugsatfda_docs/label/2019/213051s000lbl.pdf.

Sanders, M., Bruin, R., & Tran, C. (2017). Pressurised metered dose inhalers (pMDIs) with integrated flow rate guidance. *European Respiratory Journal, 50*, PA529. Available from https://doi.org/10.1183/1393003.congress-2017.PA529.

Santus, P., Ferrando, M., Baiardini, I., Radovanovic, D., Fattori, A., & Braido, F. (2019). Patients beliefs on intravenous and subcutaneous routes of administration of biologics for severe asthma treatment: A cross-sectional observational survey study. *World Allergy Organization Journal, 12*(4), 100030. Available from https://doi.org/10.1016/j.waojou.2019.100030.

Sassalos, T. M., & Paulus, Y. M. (2019). Prefilled syringes for intravitreal drug delivery. *Clinical Ophthalmology, 13*, 701−706. Available from https://doi.org/10.2147/OPTH.S169044.

Sawicki, G. S., Chou, W., Raimundo, K., Trzaskoma, B., & Konstan, M. W. (2015). Randomized trial of efficacy and safety of dornase alfa delivered by eRapid nebulizer in cystic fibrosis patients. *Journal of Cystic Fibrosis, 14*(6), 777−783. Available from https://doi.org/10.1016/j.jcf.2015.04.003.

Scheepers, H. P. A., Beaney, A. M., Le Brun, P. P. H., Neerup Handlos, V., Schutjens, M. D. B., Walser, S., & Neef, C. (2015). Aseptic preparation of parenteral medicinal products in healthcare establishments in Europe. *European Journal of Hospital Pharmacy: Science and Practice, 23*(1), 50−53. Available from https://doi.org/10.1136/ejhpharm-2015-000709.

Scherer, T., Geller, D. E., Owyang, L., Tservistas, M., Keller, M., Boden, N., Kesser, K. C., & Shire, S. J. (2011). A technical feasibility study of dornase alfa delivery with eFlow® vibrating membrane nebulizers: Aerosol characteristics and physicochemical stability. *Journal of Pharmaceutical Sciences, 100*(1), 98−109. Available from https://doi.org/10.1002/jps.22231.

Schoellhammer, C. M., Langer, R., & Traverso, G. (2016). Of microneedles and ultrasound: Physical modes of gastrointestinal macromolecule delivery. *Tissue Barriers, 4*(2), e1150235. Available from https://doi.org/10.1080/21688370.2016.1150235.

Seah, I., Zhao, X., Lin, Q., Liu, Z., Su, S. Z. Z., Yuen, Y. S., Hunziker, W., Lingam, G., Loh, X. J., & Su, X. (2020). Use of biomaterials for sustained delivery of anti-VEGF to treat retinal diseases. *Eye, 34*, 1342−1356. Available from https://doi.org/10.1038/s41433-020-0770-y.

Shah, H. D., & Shah, M. (2011). A case of look-alike medication errors. *Indian Journal of Pharmacology, 43*(4), 482−483. Available from https://doi.org/10.4103/0253-7613.83127.

Shpilberg, O., & Jackisch, C. (2013). Subcutaneous administration of rituximab (MabThera) and trastuzumab (Herceptin) using hyaluronidase. *British Journal of Cancer, 109*(6), 1556−1561. Available from https://doi.org/10.1038/bjc.2013.371.

Škalko-Basnet, N. (2014). Biologics: The role of delivery systems in improved therapy. *Biologics: Targets and Therapy, 8*, 107−114. Available from https://doi.org/10.2147/BTT.S38387.

Solsona, M. D. E., Boquet, E. M., Estruch, B. C., & Andrés, J. L. P. (2017). Impact of adherence on subcutaneous interferon beta-1a effectiveness administered by Rebismart® in patients with multiple sclerosis. *Patient Preference and Adherence, 11*, 415–421. Available from https://doi.org/10.2147/PPA.S127508.

Stadler, S. L., & Cook, T. J. (2017). PCSK9 inhibitors and managing cost in the managed care setting. *The American Journal of Managed Care, 23*(9), S149–S155. Available from https://cdn.sanity.io/files/0vv8moc6/ajmc/854a1b37d62fadfb166d2d29b279dbb706-d9aec8.pdf.

Stegemann, S. (2018). Patient centric drug product design in modern drug delivery as an opportunity to increase safety and effectiveness. *Expert Opinion on Drug Delivery, 15*(6), 619–627. Available from https://doi.org/10.1080/17425247.2018.1472571.

Stroetmann, K. A. (2015). Patient-centric care and chronic disease management: A stakeholder perspective. *Studies in Health Technology and Informatics, 208*, 324–330. Available from https://pubmed.ncbi.nlm.nih.gov/25676996/.

Stroh, M., Winter, H., Marchand, M., Claret, L., Eppler, S., Ruppel, J., Abidoye, O., Teng, S. L., Lin, W. T., Dayog, S., Bruno, R., Jin, J., & Girish, S. (2017). Clinical pharmacokinetics and pharmacodynamics of atezolizumab in metastatic urothelial carcinoma. *Clinical Pharmacology and Therapeutics, 102*(2), 305–312. Available from https://doi.org/10.1002/cpt.587.

Subhi, Y., Kjer, B., & Munch, I. C. (2016). Prefilled syringes for intravitreal injection reduce preparation time. *Danish Medical Journal, 63*(4), A5214. Available from https://ugeskriftet.dk/files/scientific_article_files/2018-11/a5214.pdf.

Subrizi, A., del Amo, E., Korzhikov-Vlakh, V., Tennikova, T., Ruponen, M., & Urtti, A. (2019). Design principles of ocular drug delivery systems: Importance of drug payload, release rate, and material properties. *Drug Discovery Today, 24*(8), 1446–1457. Available from https://doi.org/10.1016/j.drudis.2019.02.001.

Suvikas-Peltonen, E., Palmgren, J., Häggman, V., Celikkayalar, E., Manninen, R., & Airaksinen, M. (2017). Auditing safety of compounding and reconstituting of intravenous medicines on hospital wards in Finland. *International Journal of Pharmaceutical Compounding, 21*(6), 518–529. Available from https://ijpc.com/Abstracts/Abstract.cfm?ABS = 4399.

Tan, A. R., Im, S.-A., Mattar, A., Colomer, R., Stroyakovskii, D., Nowecki, Z., De Laurentiis, M., Pierga, J.-Y., Jung, K. H., Schem, C., Hogea, A., Badovinac Crnjevic, T., Heeson, S., Shivhare, M., Kirschbrown, W. P., Restuccia, E., & Jackisch, C. (2021). Fixed-dose combination of pertuzumab and trastuzumab for subcutaneous injection plus chemotherapy in HER2-positive early breast cancer (FeDeriCa): A randomised, open-label, multicentre, non-inferiority, phase 3 study. *The Lancet Oncology, 22*(1), 85–97. Available from https://doi.org/10.1016/s1470-2045(20)30536-2.

Teixeira, M. I., Lopes, C. M., Amaral, M. H., & Costa, P. C. (2020). Current insights on lipid nanocarrier-assisted drug delivery in the treatment of neurodegenerative diseases. *European Journal of Pharmaceutics and Biopharmaceutics, 149*, 192–217. Available from https://doi.org/10.1016/j.ejpb.2020.01.005.

Xeris Pharmaceuticals. (2020). The Xeris solution for conventional formulation limitations. https://www.xerispharma.com/research-development/technology.

Torres, J. A., Fischer, N. A., Seibold, L. K., Pantcheva, M. B., Kahook, M. Y., & SooHoo, J. (2018). A survey of patient perceptions and preferences for glaucoma treatment with intravitreal injections. *Ophthalmology Glaucoma, 1*(1), 61–65. Available from https://doi.org/10.1016/j.ogla.2018.06.005.

Turner, M. R., & Balu-Iyer, S. V. (2018). Challenges and opportunities for the subcutaneous delivery of therapeutic proteins. *Journal of Pharmaceutical Sciences, 107*(5), 1247–1260. Available from https://doi.org/10.1016/j.xphs.2018.01.007.

Tuvia, S., Pelled, D., Marom, K., Salama, P., Levin-Arama, M., Karmeli, I., Idelson, G. H., Landau, I., & Mamluk, R. (2014). A novel suspension formulation enhances intestinal absorption of macromolecules via transient and reversible transport mechanisms. *Pharmaceutical Research, 31*(8), 2010–2021. Available from https://doi.org/10.1007/s11095-014-1303-9.

Twarog, C., Fattah, S., Heade, J., Maher, S., Fattal, E., & Brayden, D. J. (2019). Intestinal permeation enhancers for oral delivery of macromolecules: A comparison between salcaprozate sodium (SNAC) and sodium caprate (C10). *Pharmaceutics, 11*(2), 78. Available from https://doi.org/10.3390/pharmaceutics11020078.

Ugurlu, T., & Ozaydin, T. (2014). An overview on fixed dose combinations. *Asian Journal of Pharmaceutical Technology and Innovation, 2*(9), 75–81. Available from https://www.asianpharmtech.com/articles/an-overvew-on-fxed-dosecombinations.pdf.

van den Bemt, B. J. F., Gettings, L., Domańska, B., Bruggraber, R., Mountian, I., & Kristensen, L. E. (2019). A portfolio of biologic self-injection devices in rheumatology: How patient involvement in device design can improve treatment experience. *Drug Delivery, 26*(1), 384–392. Available from https://doi.org/10.1080/10717544.2019.1587043.

Vermeire, S., D'heygere, F., Nakad, A., Franchimont, D., Fontai, F., Louis, E., Van Hootegem, P., Dewit, O., Lambrecht, G., Strubbe, B., & Baert, F. (2018). Preference for a prefilled syringe or an auto-injection device for delivering golimumab in patients with moderate-to-severe ulcerative colitis: A randomized crossover study. *Patient Preference and Adherence, 12*, 1193–1202. Available from https://doi.org/10.2147/PPA.S154181.

Vllasaliu, D., Thanou, M., Stolnik, S., & Fowler, R. (2018). Recent advances in the oral delivery of biologics: Nanomedicine and physical modes of delivery. *Expert Opinion on Drug Delivery, 15*(8), 759–770. Available from https://doi.org/10.1080/17425247.2018.1504017.

Wang, D. D., Zhang, S., Zhao, H., Men, A. Y., & Parivar, K. (2009). Fixed dosing versus body size-based dosing of monoclonal antibodies in adult clinical trials. *Journal of Clinical Pharmacology, 49*(9), 1012–1024. Available from https://doi.org/10.1177/0091270009337512.

White, S., Bennett, D. B., Cheu, S., Conley, P. W., Guzek, D. B., Gray, S., Howard, J., Malcolmson, R., Parker, J. M., Roberts, P., Sadrzadeh, N., Schumacher, J. D., Seshadri, S., Sluggett, G. W., Stevenson, C. L., & Harper, N. J. (2005). Exubera^{TM}: Pharmaceutical development of a novel product for pulmonary delivery of insulin. *Diabetes Technology and Therapeutics*, 7(6), 896–906. Available from https://doi.org/10.1089/dia.2005.7.896.

Williams, E. L., & Edwards, C. J. (2006). Patient preferences in choosing anti-TNF therapies-R1. *Rheumatology (Oxford), 45*(12), 1575–1576. Available from https://doi.org/10.1093/rheumatology/kel369.

Wynne, C. J., Ellis-Pegler, R. B., Waaka, D. S., Schwabe, C., Lehle, M., Heinzmann, D., Mangat, R., Li, C., Dick, A. E., Cranshaw, N. A., & Lum, B. L. (2013). Comparative pharmacokinetics of subcutaneous trastuzumab administered via handheld syringe or proprietary single-use injection device in healthy males. *Cancer Chemotherapy and Pharmacology, 72*(5), 1079–1087. Available from https://doi.org/10.1007/s00280-013-2273-z.

Yang, B. B., Morrow, P. K., Wu, X., Moxness, M., & Padhi, D. (2015). Comparison of pharmacokinetics and safety of pegfilgrastim administered by two delivery methods: Onbody injector and manual injection with a prefilled syringe. *Cancer Chemotherapy and Pharmacology, 75*(6), 1199–1206. Available from https://doi.org/10.1007/s00280-015-2731-x.

Yun, Y., Cho, Y. W., & Park, K. (2013). Nanoparticles for oral delivery: Targeted nanoparticles with peptidic ligands for oral protein delivery. *Advanced Drug Delivery Reviews, 65*(6), 822–832. Available from https://doi.org/10.1016/j.addr.2012.10.007.

Zhao, X., Suryawanshi, S., Hruska, M., Feng, Y., Wang, X., Shen, J., Vezina, H. E., McHenry, M. B., Waxman, I. M., Achanta, A., Bello, A., Roy, A., & Agrawal, S. (2017). Assessment of nivolumab benefit-risk profile of a 240-mg flat dose relative to a 3-mg/kg dosing regimen in patients with advanced tumors. *Annals of Oncology*, *28*(8), 2002–2008. Available from https://doi.org/10.1093/annonc/mdx235.

Ziemssen, T., Sylverster, L., Rametta, M., & Ross, A. P. (2015). Patient satisfaction with the new interferon beta-1b autoinjector (BETACONNECT™). *Neurology and Therapy*, *4*(2), 125–136. Available from https://doi.org/10.1007/s40120-015-0036-y.

Zisser, H. C. (2010). The omnipod insulin management system: The latest innovation in insulin pump therapy. *Diabetes Therapy*, *1*(1), 10–24. Available from https://doi.org/10.1007/s13300-010-0004-6.

CHAPTER 2

Clinical development of subcutaneous dosing alternatives to established intravenous formulations for monoclonal antibodies—pharmacokinetic-based bridging approach

Introduction

To enable timely access to new biotherapeutics, at initial launch, the product is commonly not yet fully optimized with regard to its drug delivery profile and product presentation. This is primarily to avoid any delay in the availability of the product related to the development of novel formulation and device technologies. Moreover, changes in treatment management, such as an increasing acceptance and demand for a flexible care setting, including home- and patient self-administration, evolve over time, thus mandating simple and ready-to-use drug delivery modalities as a lifecycle management activity. In this context, to reduce dosing complexity and healthcare capacity demands inherent to parenteral administration, lifecycle management for biological molecules often includes the development of a subcutaneous formulation as an alternative to intravenous infusions (refer to Chapter 5, Formulation and device lifecycle management of biotherapeutics—leveraging synergies across therapeutic areas).

Clinical development programs for subcutaneous formulations can be based on the established pharmacokinetic, efficacy, and safety profile with the initially approved intravenous dosing regimen. This available database is a prerequisite for applying a so-called pharmacokinetic-based bridging approach. The approach was first developed and successfully applied to the subcutaneous formulations for the two monoclonal antibodies (mAbs), trastuzumab for the treatment of human epidermal growth factor receptor

Formulation and Device Lifecycle Management of Biotherapeutics
DOI: https://doi.org/10.1016/B978-0-12-823741-0.00002-6
© 2022 Elsevier Inc.
All rights reserved.

2 (HER2)-positive breast cancer and rituximab for the treatment of non-Hodgkin's lymphoma (NHL) and chronic lymphocytic leukemia (CLL).

The subcutaneous dosing option of trastuzumab is provided in a single-use vial containing 600 mg trastuzumab and 10,000 units hyaluronidase in a 5-milliliter (mL) solution. The formulation was approved in 2013 by the European Medicines Agency (EMA) for the treatment of patients with HER2-positive metastatic breast cancer (mBC) and early breast cancer (eBC) (Herceptin; EMA, 2013), and in 2019 by the United States Food and Drug Administration (FDA) for the treatment of HER2-overexpressing breast cancer (Herceptin Hylecta; FDA Label, 2021). Subcutaneous rituximab is provided in a single-use vial containing 1400 mg rituximab and 23,400 U hyaluronidase in a 11.7-mL solution and received initial European marketing authorization in 2014 for the treatment of NHL (MabThera; EMA, 2014c), followed by the approval of an additional dose strength of 1600 mg rituximab and 26,800 U hyaluronidase in a 13.4-mL solution for the treatment of CLL in 2016 (MabThera; EMA, 2016). The product was subsequently approved by the FDA in 2017 for the treatment of follicular lymphoma (FL), diffuse large B-cell lymphoma, and CLL for patients who have received at least one full dose of a rituximab product by intravenous infusion (Rituxan Hycela; FDA Label, 2021).

Today, pharmacokinetic-based bridging programs for subcutaneous formulations of mAbs demonstrating pharmacokinetic noninferiority versus intravenous administration and providing supporting safety, immunogenicity, and efficacy data have become an established approach and are applied for biotherapeutics that are either already approved or in late-stage clinical development (Tan et al., 2021; Mateos et al., 2020). Data generated in one indication could be extrapolated to additional indications when applying the same dose and dosing regimen (Bittner et al., 2014). This development pathway forms the basis for accelerating formulation and device lifecycle management and may allow earlier access to more customer- and patient-centric product presentations.

This chapter describes the scientific rationale and application of clinical bridging approaches in the development of subcutaneous formulations for mAbs on the basis of similarities and differences for novel formulation developments for small molecules. Discrepancies between the respective indications, the patient population, and the overall tolerability profile of an mAb are highlighted with the aim to guide researchers on defining the most appropriate development pathway for their products.

Clinical bridging approaches from established intravenous to subcutaneous administration

Clinical bridging approach—definition

The focus in this chapter is on formulation lifecycle management of biologics that are in late-stage development or already approved and that encompass a change in the route of administration from intravenous to subcutaneous infusion, while keeping the dosing frequency unchanged.

The underlying clinical development approaches for novel subcutaneous formulations are designed to "bridge" to the data obtained with the intravenous formulation. Thus assumptions on pharmacokinetic–efficacy correlations of the respective molecules for the new subcutaneous route are based on the available data sets previously generated with the intravenous route. Related clinical development pathways will therefore be referred to as "clinical bridging approach" throughout the text.

Clinical bridging approaches for new formulations for oral small molecule drugs—the precedent

Formulation lifecycle management with focus on improving adherence and compliance with the predominantly oral dosing regimens has been initially introduced for small molecules. Key challenges in the area of oral delivery of small molecules represent polypharmacy, commonly defined as using from 5 to 10 prescription drugs (Golchin et al., 2015), and the so-called food effect, requiring to take certain medications in the fasted stage while others have to be taken with food (Deng et al., 2017). These factors have been associated with poor compliance with the treatment plan and can result in adverse outcomes, including mortality, falls, adverse drug reactions, and increased length of stay in the hospital (Masnoon et al., 2017).

A *meta*-analysis across different acute and chronic disease states suggested that a reduction in both, dosing frequency and the number of tablets or capsules a patient has to take every day, may improve adherence to therapies and thus decrease healthcare costs associated with the above consequences of nonadherence (Srivastava et al., 2013). Common approaches to reduce the significant medication burden placed on patients represent the development of long-acting or slow release technologies (Altreuter et al., 2018), as well as fixed-dose combinations with two or more molecules coformulated within the same dosage form (Schernthaner, 2010).

Global health authorities have issued guidance documents detailing the required technical and clinical database to support formulation changes for small molecules. In the event of postapproval changes in formulation excipients, batch sizes, or manufacturing equipment and processes that do not impact the dose and dosing frequency, the FDA issued "Guidance for Industry" documents for both "immediate release solid oral dosage forms" (SUPAC-IR, 1995) and "modified-release solid oral dosage forms" (SUPAC-Mr, 1997). Here, depending on the level of change in the new solid oral dosage forms, equivalence to the reference product could either be established via dissolution profiles or may require the demonstration of bioequivalence in drug exposures measured as area under the curve (*AUC*) and maximum serum concentrations (C_{max}). Similar guidance is available from the EMA that describes the option for performing in vitro tests instead of in vivo studies (EMA, 2017b) for products that fulfill the criteria for a Biopharmaceutics Classification System-based biowaiver (EMA, 2020).

Separate regulatory guidance documents have been developed for a change from an established immediate release to a modified or sustained release oral dosage form (EMA, 2014b). Due to the slower release from the formulation in the gastrointestinal tract as compared to an immediate release formulation, sustained release formulations are applied to reduce the dosing frequency and pharmacokinetic peak-trough fluctuations. With the aim of achieving comparable overall exposure, individual dose levels of corresponding controlled release formulations can differ from those of their immediate release counterparts (Hsu et al., 2015; Wheless & Phelps, 2018). These differences are reflected in the above EMA "Guideline on the pharmacokinetic and clinical evaluation of modified release dosage forms." In the case of an "application for a modified release formulation of a drug that is authorised in a formulation with a different release rate (e.g., immediate release formulation...)" that aims at reaching "... a similar total exposure (*AUC*) to active substance as for the immediate release formulation...," "... pharmacokinetic data alone may not be sufficient for evaluating whether the benefit/risk ratio of the modified release formulation is comparable to the corresponding doses of the immediate release form...." The EMA guidance further specifies that as "... a principle, comparative clinical efficacy and safety data are needed in addition to pharmacokinetic data for modified release products developed after the immediate release formulation, unless adequately justified," in exceptional cases, "clinical trials may be considered unnecessary" based on "a well-defined relationship between plasma concentration(s) of the active substance/active metabolite(s) and clinical response. In this case the same or a better level of efficacy and safety has to be concluded from PK [pharmacokinetic]/PD [pharmacodynamic] studies."

Similarly, the FDA Guidance for Industry "Exposure-response relationships - study design, data analysis, and regulatory applications" states that "... in some cases, exposure-response data can support the approval of a new drug delivery system (e.g., a modified-release dosage form) when the PK [pharmacokinetic] profile is changed intentionally relative to an approved product, generally an immediate release dosage form. A known exposure-response relationship could be used to determine the clinical significance of the observed differences in exposure, and to determine whether additional clinical efficacy and/or safety data are recommended" (FDA, 2003).

FDA and EMA guidelines on the development of fixed-dose combinations for molecules that are established for combination therapy in separate formulations follow the same basic concept (EMA, 2017a; FDA, 2006). For fixed-dose combinations developed for substitution therapy in "patients already stabilised on optimal doses of the combination of the same, but separately administered, active substances, taken at the same dose interval and time" (EMA, 2017a), manufacturers have to provide sufficient justification on the relevant contribution to the desired therapeutic effect. Under the assumption that (1) the efficacy and safety profile of combination therapy has been established and (2) the proposed fixed-dose combination will be administered with the same dose and dosing frequency as previous concomitant administration of the single actives in different formulations, "demonstration of similar pharmacokinetics of the fixed combination medicinal product vs its individual active substances taken simultaneously" is required (EMA, 2017a). The "rate and extent of absorption of each component in the FDC [fixed-dose combination] have to be the same as those of each component" (FDA, 2006).

These bridging concepts to novel oral formulations for small molecules have been applied since decades and are based on an existing drug exposure correlation to efficacy and safety outcomes.

Clinical bridging approaches from established intravenous to subcutaneous administration—the underlying concept

When conceptualizing the development programs for the subcutaneous dosing alternatives for the mAbs trastuzumab and rituximab, the basic hypotheses supporting the approval of novel oral formulations of small molecules did serve as a precedent. Precedent-based hypothesis building requires an understanding of the similarities and differences between the prior example and the new product in development. Based on these considerations, the key questions that needed to be answered with the development program had been defined.

The underlying assumption was that similar to small molecules, also for mAbs, exposure could be used as a surrogate for efficacy and safety and consequently the same principle of using a pharmacokinetic-based bridging approach could be applied for the development of new formulations of mAbs (Bittner et al., 2018). In analogy to the regulatory guidelines for small molecules, the described fundamental development concept was expected to be applied independent of the product and indication. Molecule- and patient population-specific requirements had to be assessed on a case-by-case basis.

As absorption of mAbs from the subcutaneous tissue into the systemic circulation occurs predominantly via the lymphatic system (McLennan et al., 2005), specific focus was on assessing any impact of the administration route change on the immunogenicity profile of the molecules. Thus in line with the EMA's "Guideline on immunogenicity assessment of biotechnology-derived therapeutic proteins," 1-year immunogenicity data had to be available prior to market authorization (EMA, 2008). Moreover, an assessment of the possible impact of antidrug antibodies on efficacy and safety, data in individual patients was preplanned in the clinical trial protocols.

These considerations are also reflected in the regulatory guidelines on the development of biosimilars. According to the FDA, "...a biosimilar is a biological product that is highly similar to and has no clinically meaningful differences from an existing FDA-approved reference product" (FDA, 2017) and the EMA states that "...a biosimilar is a biological medicinal product that contains a version of the active substance of an already authorised original biological medicinal product..." (EMA, 2014a). For generics that are identical copies of original small molecule drugs, demonstration of bioequivalence to the reference product is usually sufficient for a marketing authorization in all therapeutic indications the reference product is approved for (Curigliano et al., 2016). The development program for biosimilars, however, consists of additional components. Here, the required clinical data package has to be designed to demonstrate comparable pharmacokinetics, pharmacodynamics, and immunogenicity.

Moreover, at least one phase 3 clinical comparability trial is required to confirm similar efficacy and safety to the originator product in a population that is sufficiently sensitive to detect any differences between the products (Agbogbo et al., 2019; Curigliano et al., 2016).

Clinical bridging approaches from established intravenous to subcutaneous administration—special considerations

Fixed dose versus body size-adjusted dose

To draw on the full potential of a subcutaneous dosing alternative, the biological molecule is ideally dosed with a fixed dose independent of a patient's body weight. This regimen omits the need for dose calculation and preparation steps and facilitates dosing in a flexible care setting with ready-to-use injection devices, such as prefilled syringes, autoinjectors, or wearable on-body/patch injectors (Bittner & Schmidt, 2021).

Traditionally, to reduce the variability in drug exposure and thus variation between individuals in response to chemotherapy, dose levels were adjusted based on an individual patient's body weight or body surface area (Reilly & Workman, 1993). Especially for antineoplastic agents with a narrow therapeutic window, body size-based dosing follows the aim of achieving a therapeutic response while maintaining an acceptable safety profile (Egorin, 2003). This dosing approach however, comes with a number of disadvantages including dose-calculation errors or wastage, if only part of the product in a single-use vial is required (Hall et al., 2020).

For molecules with a wider therapeutic window, the need for highly precise dosing is far less essential as compared to cytotoxic small molecules (Egorin, 2003). This is the case, for example, for many mAbs that exhibit a relatively large therapeutic window due to their target specificity. Here, body size contribution to pharmacokinetic variability is typically low (Bai et al., 2012). Moreover, body size-adjusted dosing for mAbs does not take into account other factors relevant for the clearance from the systemic circulation, such as the extent of the patient-specific target expression, target-mediated clearance, or the binding affinity to the target (Ryman & Meibohm, 2017).

To address the fundamental question about the selection of the most appropriate dosing principle, Wang et al. conducted a study comparing the pharmacokinetic and pharmacodynamic variability for 12 mAbs either administered with a fixed dose or a body size-adjusted dose. The results revealed a similar performance of the two principle dosing approaches for the investigated mAbs. It was found that the intersubject variability of exposure was lower following fixed dosing for some antibodies and body size-based dosing for others. The recommendation from this study was to apply fixed dosing in first-in-human studies in adults, because fixed dosing or dosing based on body weight categories eases dose preparation, lowers cost, and reduces the risk of dosing errors (Wang et al., 2009).

Fig. 2.1 is a simplified schematic representation of the impact of different dosing principles on drug exposure depending on the patient's body weight. The basic concept behind the schematic is that body weight-adjusted dosing of a mAb may overadjust for the body size effect, as the pharmacokinetics are rarely proportional to body size (Bai et al., 2012). Under this assumption, following dosing with a fixed dose, mean exposure of a mAb is expected to be higher in patients with a lower body weight compared to heavier patients. If body size- or body weight-adjusted dosing is applied, the situation is the reverse. In this situation, the exposure in heavier patients would be higher compared to lighter patients. The use of fixed-dose bands accounting for different body weight categories could be a dosing approach for biological molecules with a smaller therapeutic window (Hendrikx et al., 2017).

Today, the advantages of fixed dosing regimens are broadly recognized and have, for example, led to postapproval dosing optimizations for the cancer immunotherapy treatments pembrolizumab (Keytruda; FDA Label, 2021) and nivolumab (Opdivo; FDA Label, 2021). Both mAbs were initially developed with a body weight-adjusted dosing regimen. Supported by pharmacokinetic modeling and simulation analyses (Ogungbenro et al., 2018), the treatments are now also available with an alternative fixed dosing regimen. In either case, the above-described body weight-dependent

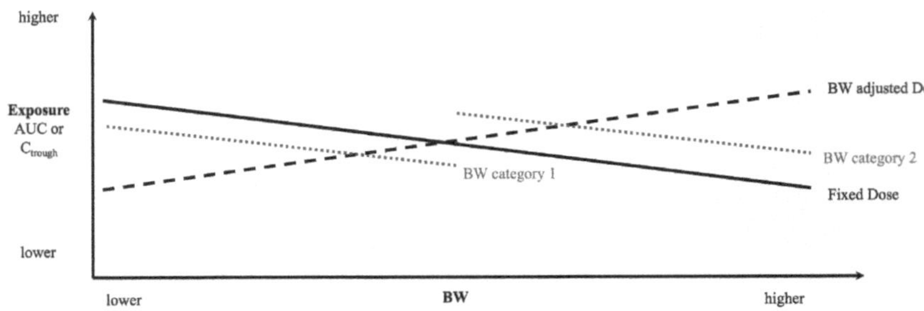

Figure 2.1 Simplified schematic representation of the impact of different dosing principles on drug exposure depending on the patient's body weight. *AUC*, area under the serum concentration-time curve; *BW*, body weight; C_{max}, maximum serum concentrations; C_{trough}, serum trough concentrations; *IV*, intravenous; *SC*, subcutaneous.

exposure profile was observed. At an overall similar exposure distribution, with the body weight-adjusted dosing regimen, lower weight patients tended to have lower antibody exposures relative to higher weight patients, in contrast to fixed dosing (Freshwater et al., 2017; Zhao et al., 2017).

Taking the above elaborations into account, biotherapeutics should be developed using a fixed dosing approach where feasible (Bai et al., 2012; Wang et al., 2009), considering the observed pharmacokinetic variability, the biodistribution profile, the therapeutic window, as well as the specific patient population (i.e., pediatrics, elderly) that may mandate the selection of the most appropriate dosing concept on a case-by-case basis.

From the approved mAbs in oncology that are offered with intravenous and subcutaneous dosing alternatives, intravenous versions of trastuzumab, rituximab, and daratumumab are dosed using a body size-dependent regimen (Herceptin, MabThera, Darzalex; Electronic Medicines Compendium, 2021), while intravenous pertuzumab is available with a fixed dose both for the intravenous loading and maintenance doses (Perjeta; Electronic Medicines Compendium, 2021). All subcutaneous formulations are administered with a fixed dose (Herceptin, MabThera, Darzalex, and Phesgo; Electronic Medicines Compendium, 2021).

Impact of changes in dosing frequency on the development program for the subcutaneous route

Less frequent dosing is typically preferred by patients and especially when it comes to parenteral drug administration, a reduction of in-clinic dosing days has been demonstrated to avoid hospital resource utilization (Boccara et al., 2017; Wai et al., 2000). When planning development of a new subcutaneous formulation for a biological drug as a lifecycle management activity, maintaining the dosing frequency of the intravenous regimen is a critical factor for the applicability of a pharmacokinetic-based bridging program.

The dosing frequency with intravenous infusions during which the diluted drug formulation is slowly administered via an infusion bag is not impacted by the total dose of the biopharmaceutical. For instance, immunoglobulins for the treatment of primary immune deficiency are administered at individual intravenous doses as high as 200–800 mg/kg (Mahmood et al., 2020). By contrast, high-dose subcutaneous formulations typically result in comparatively high dosing volumes that may mandate drug administration at shorter dosing intervals. Consequently, in the past, subcutaneous dosing alternatives for high-dose biotherapeutics were typically administered more frequently compared to their intravenous counterparts (Jolles et al., 2011).

Increasing the dosing frequency to account for a high subcutaneous injection volume, for example, from monthly (q1m) intravenous dosing to weekly (q1w) subcutaneous administration, would require a larger development program including adequately powered efficacy assessments in larger phase 3 trials. One example is tocilizumab that introduced a subcutaneous formulation administered q1w or every 2 weeks (q2w) as an alternative to the marketed intravenous formulation dosed every 4 weeks (q4w). The underlying development program did consist of two phase 3 studies. The SUMMACTA study in patients with moderate-to-severe rheumatoid arthritis compared intravenous tocilizumab dosed at 8 mg/kg q4w to subcutaneous tocilizumab at 162 mg with a q1w regimen and enrolled 1262 patients to demonstrate noninferiority in the primary efficacy endpoint ACR20 scores (Burmester et al., 2014). The BREVACTA study tested subcutaneous tocilizumab at 162 mg injected q2w and enrolled 656 patients with rheumatoid arthritis who had an inadequate response to disease-modifying antirheumatic drugs to show superiority in ACR20 versus placebo (Kivitz et al., 2014).

Approved precedents of programs for mAbs in oncology that introduced a new subcutaneous formulation as lifecycle management and maintained the established intravenous dosing frequency by developing a high-volume subcutaneous formulation include trastuzumab (Ismael et al., 2012), rituximab (Davies, Berge, et al., 2017; Davies, Merli, et al., 2017), the fixed-dose combination of pertuzumab and trastuzumab (Tan et al., 2021), and daratumumab (Mateos et al., 2020). The clinical development programs did consist of a pharmacokinetic-based dose finding and confirmation approach, complemented with the generation of supporting efficacy, safety, and immunogenicity data. The development pathways for the first programs that implemented this pharmacokinetic-based bridging approach, that is, subcutaneous trastuzumab and rituximab, are described in more detail below.

Efficacy, safety, and immunogenicity

Demonstration of clinical efficacy, safety, and immunogenicity is an essential aspect in the development of new biotherapies and subsequent lifecycle management. Especially, immunogenicity can result in anaphylactic reactions or severe immuno-toxicological reactions and could impair efficacy via generation of antidrug antibodies that may neutralize the biological activity of the biotherapeutic product (Brennan et al., 2010). FDA and EMA have issued guidelines outlining the immunogenicity assessments required for novel therapeutic protein products and biosimilars (EMA, 2008; FDA, 2014).

Accordingly, safety and immunogenicity are important assessments to complement any pharmacokinetic-based bridging program for new formulations and product presentations of biotherapeutics. This has implications for the selection of the patient population for the supporting clinical trials. Although a new subcutaneous formulation of an established biotherapeutic does not represent a biosimilar version on itself (Zharkov et al., 2019), the relevant EMA and FDA biosimilar guidances (EMA, 2012; FDA, 2019) should be considered when selecting the indication and patient population for the clinical development pathway, especially if a biological medication is approved for several indications. Namely, the guidance "Clinical pharmacology data to support a demonstration of biosimilarity to a reference product" (FDA, 2016) states that the "...study population selected should be the most informative for detecting and evaluating differences in PK [pharmacokinetic] and PD [pharmacodynamic] profiles between the proposed biosimilar product and the reference product." According to EMA's "Guideline on similar biological medicinal products containing mAbs—non-clinical and clinical issues" (EMA, 2012), a "... comparative pharmacokinetic study in a sufficiently sensitive and homogeneous study population (healthy volunteers or patients) normally forms an initial step of biosimilar mAb development." It is further recommended that "to demonstrate similar clinical efficacy and safety compared to the reference medicinal product," "the most sensitive model and study conditions (pharmacodynamic or clinical) should be used in a homogeneous patient population."

While these biosimilar guidelines were not yet available at the time of defining the clinical bridging approach for subcutaneous trastuzumab and rituximab, the underlying development programs followed a similar pathway (Jackisch et al., 2015; Kolberg et al., 2019). Pharmacokinetic-based dose finding was conducted in subjects with comparatively low expected variability in exposure. Namely, initial studies were conducted in healthy volunteers and patients with HER2-positive eBC for trastuzumab and in patients with FL in the maintenance setting who had previously responded to induction treatment for rituximab. Data on clinical efficacy, safety, and immunogenicity of subcutaneous versus intravenous trastuzumab were subsequently generated in patients with HER2-positive eBC, a population who was considered more sensitive as compared to patients with HER2-positive mBC concomitantly receiving immunosuppressive treatments, such as chemotherapy and radiation. Coprimary endpoints were C_{trough} prior to surgery and pathological complete response (pCR) (Ismael et al., 2012). For rituximab, pharmacokinetics of the selected dose level was confirmed in patients with FL in the maintenance setting. Additional pharmacokinetic and efficacy data [overall response rate (ORR)] were collected at the end of induction treatment to reduce the impact of target-mediated clearance on pharmacokinetic variability (Davies, Berge, et al., 2017; Davies, Merli, et al., 2017) (see below for details). Samples analyzed for the formation of antidrug antibodies were taken at different timepoints throughout all studies and in case found positive, assessed for a possible correlation with exposure, efficacy, and safety at an individual patient level.

Patient and healthcare provider preferences and health economic impact

Today, the relevance of patient and healthcare provider preferences in the design of customer-centric treatments is generally acknowledged. This is particularly important, as enhanced satisfaction of the demands of the target population has been shown to be associated with higher adherence, persistence, and better treatment outcomes (van Overbeeke et al., 2019). Moreover, to understand the impact of a change in the dosing regimen on healthcare capabilities, it is mandated to assess the overall costs and resources attributed to the novel drug delivery modality (Rule et al., 2014). Consequently, for decisions linked to lifecycle management of a biotherapeutic agent, such as the development of new formulations, administration routes, or injection devices, insights into patient and healthcare provider needs and into the health economic impact have to be generated early on. Also, learnings from other products with similar drug delivery improvements should be taken into account in this context.

Patient and healthcare provider preferences are, for example, assessed during clinical development as integral part of clinical studies (Blackwood et al., 2020). In case a formulation change is envisaged following initial approval of the drug, related input can be gathered in dedicated preference trials (Ahmann et al., 2014) or in so-called time-and-motion studies that assess possible time savings in cooperation with healthcare institutions

(Burcombe et al., 2013). Furthermore, insights into how to best optimize the product presentation or the drug delivery profile can be gained through patient and healthcare provider advisory boards (Monaco et al., 2021). Such understanding on stakeholder demands has become an important parameter for both health technology assessments and regulatory data reviews in assessing the benefit–risk ratio of a new product presentation (Mott, 2018).

Dedicated preference and health economic studies have been conducted for both trastuzumab and rituximab. A detailed summary of trial designs and results was published previously (Bittner et al., 2018).

Label extension to include a flexible care setting

The main benefits associated with a patient's preference for subcutaneous formulations analyzed in a systematic review by Stoner et al. (2015) were time savings and the possibility to receive treatment at home. Accordingly, product optimizations encompassing a change in route of administration from intravenous to subcutaneous dosing ideally also come with a label extension for home-, or patient self-administration, if provided in a suitable product presentation.

Prerequisites for such label extensions are that the subcutaneously dosed biotherapeutic molecule is safe and well tolerated and that patients and/or their caregivers can reliably inject the medication without the supervision of a healthcare provider. Demonstration of adequate safety requires additional data beyond the safety data generated in a comparatively small pharmacokinetic-based bridging program.

In case the general safety profile of a biotherapeutic is largely administration route independent, manufacturers should start with a thorough analysis of the molecule's systemic adverse event profile following intravenous dosing. The overall safety profile following intravenous and subcutaneous dosing together with the tolerability of an individual patient will guide the judgment on the molecule's suitability for patient home- and self-administration. The topic will be discussed in more detail in Chapter 4, Enabling a flexible care setting in oncology—how to realize home- and self-administration of biotherapeutics in cancer care.

Selection of subcutaneous injection sites

To account for individual patient preferences, generating supporting data for the use of multiple subcutaneous injection sites, such as the abdomen, thigh, or upper arm should be the aim of any development program. This is especially important as different anatomic injection sites have been associated with a different degree of pain (Usach et al., 2019). In particular, when it comes to home- or patient self-administration, to ensure compliance with the dosing procedure, a patient should have the choice on where the medication is administered and the option to alternate between injection sites.

Following subcutaneous injection, therapeutic proteins larger than approximately 20 kDa, such as mAbs, are predominantly absorbed via the lymphatic system prior to reaching the blood capillaries (Richter & Jacobsen, 2014; Viola et al., 2018). Contingent upon the injection site, biotherapeutics are transported through different lymphatic vessels and nodes before entering systemic blood circulation. These particularities in the absorption and transport pathways may impact bioavailability and the pharmacokinetic variability due to differences in proteolysis within the lymphatic system (Varkhede & Forrest, 2018). Namely, injection site-dependent differences in the absorption rate of a large molecule may impact its C_{max} and time to reach C_{max} (t_{max}). In the event that the established correlation between pharmacokinetics and efficacy reveals that the overall exposure and/or C_{trough} rather than C_{max} drives outcomes, differences in peak concentrations following injection into varying subcutaneous areas of the body may not be clinically relevant.

For example, for benralizumab, an mAb approved as add-on maintenance treatment for patients aged 12 years and older with uncontrolled asthma with eosinophilic inflammation (Fasenra; Electronic Medicines Compendium, 2021), exposure was approximately 15%–30% greater for thigh versus abdomen or upper arm administration. This difference was considered as unlikely clinically relevant, as consistent eosinophil depletion was observed following administration independent of the injection site (Ubaldo et al., 2021) and the use of all injection sites is permitted as per label (Fasenra; Electronic Medicines Compendium, 2021).

The currently approved high-volume subcutaneous formulations for the mAbs trastuzumab, rituximab, and daratumumab (Herceptin, MabThera, Darzalex; Electronic Medicines Compendium, 2021) and the fixed-dose combination of pertuzumab and trastuzumab (Phesgo; Electronic Medicines Compendium, 2021) only permit injection into one injection site, that is, the thigh or the abdomen. This is a consequence of restricting subcutaneous administration in the clinical studies to one area only. To add additional injection sites, future trials need to generate pharmacokinetic data for different areas of the body and possible differences in

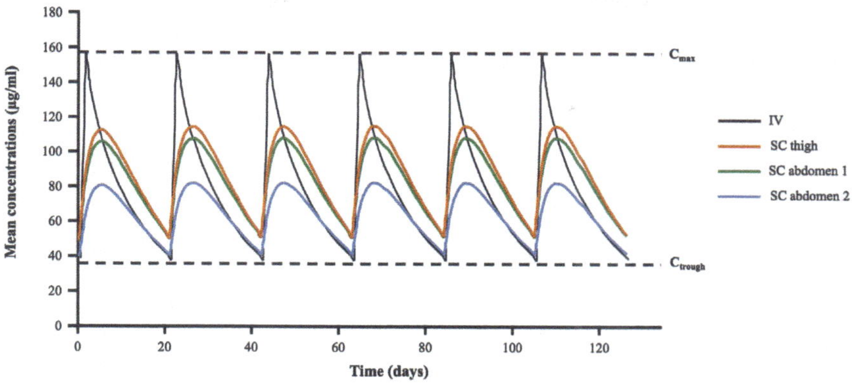

Figure 2.2 Schematic representation of IV versus SC administration at three different injection sites and simulated mean concentration-time profiles. Concentration-time profiles for SC abdomen 1 and SC abdomen 2 represent injection into the abdomen with different hypothetical mean concentrations. Considering the variability around the serum concentration of a mAb, SC abdomen 1 might still be within the variability range of exposure for the approved intravenous and subcutaneous dosing regimens. SC abdomen 2 would likely result in underexposure of a number of patients compared to the exposure range with the approved dosing regimens and injection sites. C_{max}, maximum plasma concentration, C_{trough}, serum trough concentration; *IV*, intravenous; *SC*, subcutaneous.

the pharmacokinetic profile need to be assessed for clinical relevance. Thus in case observed differences in AUC or C_{trough} with thigh versus abdominal administration would still be within the variability range of exposure for the approved intravenous and subcutaneous formulations, comparable efficacy would be expected (Fig. 2.2).

Clinical bridging programs for the subcutaneous formulations of trastuzumab and rituximab

The new formulations of trastuzumab and rituximab were the first large-volume subcutaneous injectables approved in oncology (Bittner et al., 2012, 2014, 2018; Shpilberg & Jackisch, 2013). The molecules are coformulated with recombinant human hyaluronidase to facilitate dispersion in the interstitial space (Bookbinder et al., 2006; Connor et al., 2020). This way the frequency of the intravenous dosing regimens could be maintained despite the comparatively large injection volumes of 5 and 11.7 mL, respectively.

The fundamental assumptions underlying the pharmacokinetic-based bridging approach were that (1) the new formulation and subcutaneous route of administration would not change the molecular properties of the active molecule and (2) comparable exposure to the intravenous dosing regimen would lead to comparable efficacy.

Trastuzumab

Subcutaneous trastuzumab—dose finding approach

The q3w subcutaneous dosing alternative for trastuzumab was developed with the goal to ensure noninferior C_{trough} to intravenous dosing and thus a comparable efficacy and safety profile. While intravenous trastuzumab is approved with a q1w and q3w dosing regimen, the subcutaneous formulation is available with the q3w schedule only. Approved intravenous dosing regimens and indications at the time of the start of subcutaneous development are summarized in Table 2.1.

Dose selection for intravenous trastuzumab was guided by a mouse xenograft model demonstrating a superior outcome against tumor growth when exceeding a trastuzumab serum concentration of 20 μg/mL (data on file, F. Hoffmann-La Roche). Intravenous trastuzumab was first developed with a q1w dosing regimen that included a loading dose of 4 mg/kg to achieve the target concentration more rapidly and a maintenance dose of 2 mg/kg. The conclusion that the q1w intravenous dosing regimen leads to the saturation of accessible target sites and therefore to optimal clinical efficacy was supported by a plateau effect in the dose–efficacy relationship (Vogel et al., 2002). Namely, comparing the approved q1w regimen with a higher q1w dose, that is, a loading dose of 8 mg/kg and q1w maintenance doses of 4 mg/kg in 114 patients with HER2-overexpressing mBC (59 patients on the standard dose and 55 patients on the higher dose) showed no apparent benefit of the higher over the lower approved dose based on the efficacy and safety endpoints.

Subsequently, intravenous trastuzumab was provided with a second less frequent 3-weekly dosing (q3w) regimen with a loading dose of 8 mg/kg and a maintenance dose of 6 mg/kg, developed to achieve the average exposure obtained with the q1w regimen (Baselga et al., 2005). A population pharmacokinetic analysis with data primarily from the mBC setting revealed that predicted median trough concentrations over a period of 3 weeks at steady state for the q1w and q3w regimens exceed the target minimum or trough concentration (C_{trough}) of 20 μg/mL in more than 90% of patients (Bittner et al., 2012).

The availability of two different dosing regimens for intravenous trastuzumab was the foundation for the pharmacokinetic-based bridging

Table 2.1 Approved trastuzumab dosing regimens at time of subcutaneous development initiation.

Regimen	Indication	Dose and dosing frequency trastuzumab	Treatment duration
H monotherapy	HER2-positive mBC for patients who have received at least two chemotherapy regimens for their metastatic disease	Loading dose of 8 mg/kg followed by 6 mg/kg maintenance dose q3w	Until disease progression
		Loading dose of 4 mg/kg followed by 2 mg/kg maintenance dose q1w	Until disease progression
	HER2-positive eBC for patients following surgery, chemotherapy and radiotherapy	Loading dose of 8 mg/kg followed by 6 mg/kg maintenance dose q3w	1 year or until disease recurrence
H chemotherapy	HER2-positive mBC for patients who have not received chemotherapy for their metastatic disease	Loading dose of 8 mg/kg followed by 6 mg/kg maintenance dose q3w	Until disease progression
		Loading dose of 4 mg/kg followed by 2 mg/kg maintenance dose q1w	Until disease progression
	HER2-positive eBC in combination with adjuvant chemotherapy consisting of docetaxel and carboplatin or in combination with neoadjuvant chemotherapy followed by adjuvant Herceptin therapy, for locally advanced (including inflammatory) disease or tumors >2 cm in diameter	Loading dose of 8 mg/kg followed by 6 mg/kg maintenance dose q3w	1 year or until disease recurrence
	HER2-positive mGC in combination with capecitabine or 5-fluorouracil and cisplatin is who have not received prior anticancer treatment for their metastatic disease	Loading dose of 8 mg/kg followed by 6 mg/kg maintenance dose q3w	Until disease progression

eBC, early breast cancer; *H*, herceptin; *mBC*, metastatic breast cancer; mGC, metastatic gastric cancer; *q1w*: every week; *q3w*: every 3 weeks.

approach to the subcutaneous dosing regimen. Namely, the correlation between pharmacokinetics and efficacy for the intravenous regimens did support the hypothesis that to achieve comparable outcomes following subcutaneous and intravenous administration, C_{trough} with either regimen needs to exceed the target concentration of 20 μg/mL. This aspect is particularly important as inherent to the change in the administration route, at equivalent overall exposure, following subcutaneous administration and subsequent slow absorption into the systemic circulation, C_{max} is lower compared to intravenous infusion. Likewise, intravenous trastuzumab peak levels are higher with q3w compared to q1w administration. Despite this difference in C_{max} between the intravenous dosing regimens, both treatments exhibit a similar efficacy and safety profile (Baselga et al., 2005). Hence, the dose for subcutaneous trastuzumab with a q3w dosing regimen was selected with the aim to achieve noninferior C_{trough} compared to the intravenous regimen to result in comparable target receptor saturation and thus comparable efficacy. C_{max} of the subcutaneous regimen was estimated to be bracketed by those of the q1w and q3w intravenous regimens (Fig. 2.3).

To further facilitate subcutaneous administration, the formulation was developed with a fixed dose for all patients independent of body weight

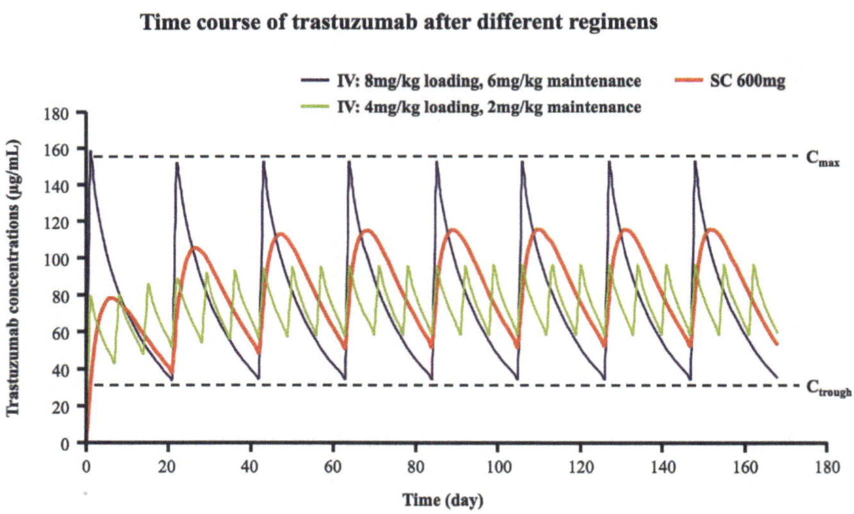

Figure 2.3 Simulated mean serum concentration–time profiles of the every week and every 3 week intravenous and the every 3 week subcutaneous dosing regimens. C_{max}, maximum serum concentrations; C_{trough}, serum trough concentrations; *IV*, intravenous; *SC*, subcutaneous.

or treatment cycle, thus also omitting the need for a loading dose. A population pharmacokinetic modeling and simulation approach was applied to predict a fixed subcutaneous dose of trastuzumab that would result in noninferior C_{trough} and AUC to the q3w intravenous regimen from the first dose onward. A subcutaneous dose of 600 mg was calculated to fulfill the preset requirements to result in trastuzumab serum exposures within the 5th percentile, median, and 95th percentile for C_{trough} in cycle 1 and cycle 7, and for the area under the serum concentration–time curve from time zero to time 21 ($AUC_{0-21\ days}$) in cycle 7 to achieve exposures that are at least as high as those with intravenous dosing (Hourcade-Potelleret et al., 2014). Fig. 2.3 illustrates the simulated mean pharmacokinetic profiles of the q1w and q3w intravenous and the q3w subcutaneous dosing regimens.

The subcutaneous formulation for trastuzumab was subsequently developed based on a pharmacokinetic-based dose finding and dose confirmation study in healthy male volunteers and patients with adjuvant HER2-positive eBC (Wynne et al., 2013) and a phase 3 noninferiority study in patients with neoadjuvant/adjuvant HER2-positive eBC, with C_{trough} and pCR as coprimary endpoints (Ismael et al., 2012). This dose selection concept has since then been applied to a number of mAbs, including trastuzumab, rituximab, daratumumab, and the fixed-dose combination of pertuzumab and trastuzumab (Davies, Berge, et al., 2017; Davies, Merli, et al., 2017; Mateos et al., 2020; Tan et al., 2021).

Subcutaneous trastuzumab—nonclinical development

The nonclinical studies supporting approval of the subcutaneous trastuzumab formulation were conducted with the aim to confirm the hypothesis that C_{max} does not contribute to efficacy of trastuzumab (refer to Fig. 2.3) and that the established safety profile of the mAb is not negatively affected with the alternative administration route.

Namely, a nonclinical xenograft study in BALB/c nu/nu mice transplanted with the high HER2-expressing nonsmall-cell lung cancer cell line Calu-3 demonstrated that regardless of the mode of injection, that is, intravenously versus subcutaneously, the same minimum serum concentrations (C_{min}) of trastuzumab resulted in a similar pharmacodynamic effect. Moreover, data revealed that higher C_{max} levels following intravenous administration did not correlate with efficacy compared to subcutaneous doses that delivered the same overall exposure (expressed as AUC) (Bittner et al., 2012).

With the change in the route of administration, additional toxicology studies had been performed. Subcutaneous local tolerability was assessed in rabbits and a 13-week repeat-dose toxicity study was conducted in cynomolgus monkeys to bridge to the studies of intravenous trastuzumab (Herceptin; EMA, 2013).

Subcutaneous trastuzumab—clinical development

The clinical development program did consist of a phase 1b dose finding/dose confirmation trial enrolling both healthy male volunteers and patients with HER2-positive eBC in the adjuvant setting (Fig. 2.4) and a phase 3 noninferiority study in patients with HER2-positive eBC (Fig. 2.5).

The phase 1b study was designed as a 2-part trial, including pharmacokinetic interim analyses and the option to adjust the dose levels for subsequent cohorts. In part 1 of the study, eight subjects were enrolled into each cohort. The initial dose finding part was conducted in healthy male volunteers to avoid possible under dosing of patients, as the subcutaneous bioavailability was unknown. Following interim analyses, an adjusted dose of 8 mg/kg was calculated to result in comparable trastuzumab C_{trough} levels compared to the 6-mg/kg intravenous control. Part 2 (dose confirmation) was conducted in two cohorts of 20 patients each. In the first

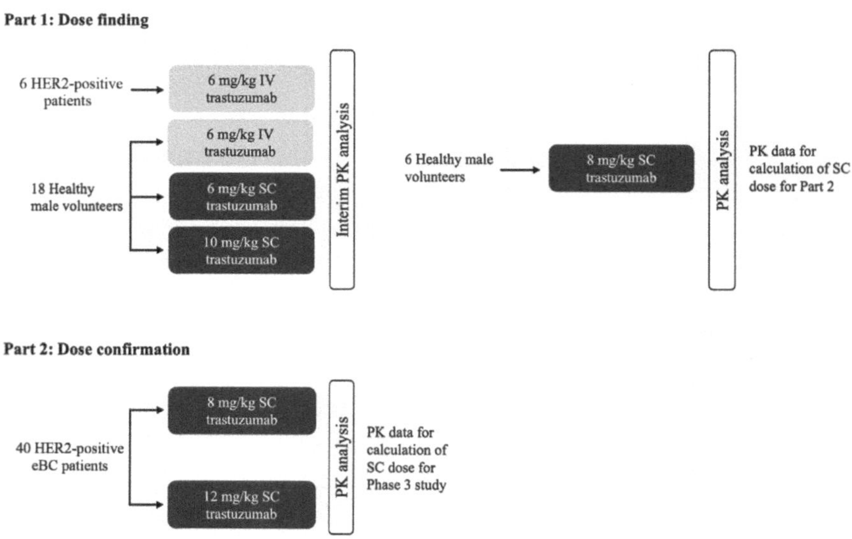

Figure 2.4 Subcutaneous trastuzumab—phase 1b dose finding/dose confirmation study. *eBC*, early breast cancer; *HER2*, human epidermal growth factor receptor 2; *IV*, intravenous; *PK*, pharmacokinetic; *SC*, subcutaneous.

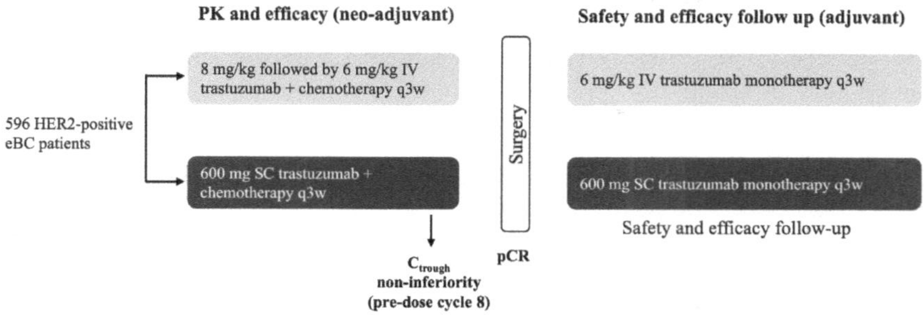

Figure 2.5 Subcutaneous trastuzumab—phase 3 noninferiority study. C_{trough}, serum trough concentration. *eBC*, early breast cancer; *HER2*, human epidermal growth factor receptor 2; *IV*, intravenous; *pCR*, pathological complete response; *q3w*, every 3 weeks; *SC*, subcutaneous.

cohort, patients received subcutaneous trastuzumab at 8 mg/kg to confirm the calculated dose from part 1 pharmacokinetic data in the target patient population. In the second cohort, a dose of 12 mg/kg was administered to assess systemic and local tolerance at higher exposures prior to start of the phase 3 study (Bittner et al., 2012). Pharmacokinetic data from this phase 1b trial were integrated into the existing population pharmacokinetic model built on the various trials previously conducted with the intravenous formulation to calculate a fixed dose to be tested in the subsequent phase 3 study (Hourcade-Potelleret et al., 2014). The confirmatory phase 3 study (HannaH; Ismael et al., 2012) was designed as an open label trial comparing trastuzumab C_{trough} and pCR as coprimary endpoints (Fig. 2.5).

HannaH was an international, multicenter study that enrolled 596 patients with HER2-positive eBC. Patients received their chemotherapy regimen (4 cycles of docetaxel at 75 mg/m^2, followed by 4 cycles of fluorouracil at 500 mg/m^2, epirubicin at 75 mg/m^2, and cyclophosphamide at 500 mg/m^2) together with either 600 mg subcutaneous trastuzumab or intravenous trastuzumab with a loading dose of 8 mg/kg followed by maintenance doses of 6 mg/kg q3w in the neoadjuvant setting until surgery. Thereafter, patients received an additional 10 cycles of subcutaneous or intravenous trastuzumab (as per initial randomization) to complete 1 year of trastuzumab therapy.

Study results demonstrated noninferiority in the coprimary endpoints C_{trough} at predose cycle 8 with 51.8 μg/mL [coefficient of variation (CV) 52.5%] in the intravenous group and 69.0 μg/mL (CV 55.8%) in the subcutaneous group. The geometric mean ratio of C_{trough} subcutaneous to

C_{trough} intravenous was 1.33 [90% confidence interval (CI), 1.24–1.44]. For the coprimary efficacy endpoint, 40.7% of patients in the intravenous group and 45.4% in the subcutaneous group achieved a pCR. The difference between the groups in pCR was 4.7% (95% CI, 4.0–13.4) demonstrating noninferiority of subcutaneous to intravenous trastuzumab for both coprimary endpoints. The incidence of grade 3–5 adverse events was similar between groups (Ismael et al., 2012). Antidrug antibodies against trastuzumab following both intravenous and subcutaneous dosing were observed transiently and were of no relevance in terms of efficacy or safety (Hegg et al., 2012).

These results confirmed the initial hypothesis that similar trastuzumab C_{trough} results in comparable efficacy independent of the route of administration and formed the basis for EMA approval of the subcutaneous formulation in patients with HER2-positive eBC. Data generated in eBC could be extrapolated to the mBC indication without the need for a dedicated study in this indication (Herceptin; EMA, 2013). The final analysis of the HannaH study with a median follow-up of approximately 6 years confirmed the efficacy and safety outcomes of the initial analysis (Jackisch et al., 2019).

Rituximab

Subcutaneous rituximab—dose finding approach

The main objective of the clinical bridging approach for the subcutaneous version of rituximab was to generate a data package that would support registration for all rituximab oncology indications and treatment regimens with the intravenous administration route. In addition, the aim was to generate further evidence in validating the pharmacokinetic-based bridging approach by demonstrating that pharmacokinetic noninferiority results in comparable clinical outcomes. The clinical development pathway followed the same principle as for subcutaneous trastuzumab, however, applying a number of adjustments to account for the different dosing regimens and frequencies as highlighted in Table 2.2.

The approved oncology indications and dosing regimens for intravenous rituximab at the time of subcutaneous development initiation are depicted in Table 2.3.

While subcutaneous trastuzumab was developed with one dose level and the least frequent intravenous dosing regimen (q3w) only, subcutaneous rituximab was made available with several doses and dosing frequencies. As different individual doses and time intervals between doses may

Table 2.2 Differences in the available intravenous dosing regimens for trastuzumab and rituximab that mandated variations in the subcutaneous development programs.

	Intravenous regimen		Subcutaneous regimen	
	Trastuzumab	Rituximab	Trastuzumab	Rituximab
Dosing frequency	q1w, q3w	q1w, q3w, q4w, q2m, q3m	q3w	q1w, q3w, q4w, q2m, q3m
Dosing principle	Body weight-adjusted	Body surface area-adjusted	Fixed dose	Fixed doses[1]
Dosing regimen	Loading and maintenance doses	Same dose(s)[1] across all cycles	Same dose across all cycles	First dose intravenous; same dose(s)[1] across all subsequent cycles

[1]Indication-dependent; q1w: every week; q3w: every 3 weeks; q4w: every 4 weeks; q2m: every 2 months; q3m: every 3 months.

impact efficacy, safety, and immunogenicity of an mAb, clinical trial data for each regimen were mandated. Deviating from intravenous trastuzumab with a body weight-based regimen, intravenous rituximab is dosed by body surface area. For trastuzumab, the fixed subcutaneous dose was calculated to omit the need for a loading dose and the antibody is administered subcutaneously from cycle 1 onward. Rituximab can exhibit infusion-related reactions especially during initial treatment cycles. These reactions are usually mild to moderate, but fatal evolutions have been reported (Paul & Cartron, 2019). It was therefore recommended at the time that the first dose should still be administered intravenously to account for possible infusion-related reactions that could be reduced in severity by slowing down the intravenous infusion rate (Paul & Cartron, 2019). If a patient tolerated the first intravenous dose, subsequent doses could be given subcutaneously.

In analogy to trastuzumab, the subcutaneous formulation for rituximab was developed with the aim to achieve comparable C_{trough} to the intravenous infusion. In the pivotal phase 3 clinical trial for intravenous rituximab in 166 patients with recurrent low-grade lymphoma treated with

Table 2.3 Approved rituximab dosing regimens at the time of subcutaneous development initiation.

Regimen	Indication	Dose and dosing frequency	Treatment duration
R monotherapy	Relapsed or refractory, low grade or follicular, CD20-positive B-cell NHL as a single agent	375 mg/m^2 q1w	4 weeks
R chemotherapy	Previously untreated adult patients with stage III–IV FL in combination with chemotherapy	375 mg/m^2 q3w	8 cycles (24 weeks)
	Maintenance therapy for the treatment of adult patients with FL responding to induction therapy	375 mg/m^2 q3m	Until PD or up to 2 years
	In combination with chemotherapy for the treatment of patients with previously untreated and relapsed/refractory CLL	Prior initiation of chemotherapy: 375 mg/m^2 Subsequent cycles: 500 mg/m^2 q4w	6 cycles
R-CHOP	Treatment of adult patients with CD20-positive DLBCL in combination with CHOP chemotherapy	375 mg/m^2 q3w	8 cycles

CHOP, cyclophosphamide, vincristine, doxorubicin, prednisone; *CLL*, chronic lymphocytic leukemia; *DLBCL*, diffuse large B-cell lymphoma; *FL*, follicular lymphoma; *PD*, progressive disease; *q2m*, every 2 months; *q3m*, every 3 months; *q1w*, every week; *q4w*, every 4 weeks; *R*, rituximab.

4 weekly infusions of rituximab, an association between clinical response and higher serum levels of the antibody both during and after treatment had been observed (Berinstein et al., 1998). Namely, at follow-up, patients who responded to treatment had higher rituximab serum

concentrations (median levels 25.4 μg/mL) compared with nonresponders (median levels 5.9 μg/mL). Today, a C_{trough} of >25 μg/mL is considered the threshold to maintain for efficacy (Golay et al., 2013).

Subcutaneous rituximab—nonclinical development

In analogy to trastuzumab, the nonclinical development pathway for rituximab was conducted to provide supporting pharmacokinetic, pharmacodynamic, and safety data prior to embarking into clinical studies. Data from a mouse xenograft model using the Z138 human mantle cell lymphoma cell line suggested that at similar rituximab C_{trough} following subcutaneous and intravenous administration, similar efficacy was achieved in this xenograft model (Bittner et al., 2014). Moreover, as described for the nonclinical program for trastuzumab, local tolerability and repeat-dose toxicity studies confirmed that the safety profile for the intravenous route was not negatively impacted following subcutaneous administration (MabThera; EMA, 2014c).

Subcutaneous rituximab—clinical development

The clinical development program for subcutaneous rituximab in NHL was composed of a dose finding/dose confirmation study in patients with FL in the maintenance setting (SparkThera; Salar et al., 2014) and a phase 3 study in patients with FL in the induction setting (Sabrina; Davies, Berge, et al., 2017; Davies, Merli, et al., 2017). For the CLL label, an indication with a different dose and dosing regimen an additional dose finding/dose confirmation study was conducted (Sawyer; Assouline et al., 2015).

The pharmacokinetic profile of rituximab in patients with NHL is characterized by a target-independent pathway that is unchanged throughout the treatment period and a target-mediated elimination that decreases toward the end of treatment related to the extent of B cell/tumor-burden at baseline (Ternant et al., 2019; Yin et al., 2010). With the aim to reduce the impact of target-mediated clearance on pharmacokinetic variability during initial subcutaneous dose selection, the first dose finding trial was conducted in patients with FL who had previously responded to induction treatment in the maintenance setting. Subcutaneous dose selection was based on the hypothesis that noninferior C_{trough} levels to those achieved with intravenous dosing will result in the same degree of receptor saturation and thus noninferior efficacy outcomes.

In stage 1 of this dose finding study (SparkThera; Salar et al., 2014), 124 eligible patients who had previously received at least one intravenous dose in the maintenance setting were dosed with either a single intravenous rituximab dose of 375 mg/m^2 or a subcutaneous rituximab dose of 375 or 625 mg/m^2, assuming a bioavailability of 100% and 60%, respectively. Following an interim analysis, an additional cohort of patients did receive an adjusted dose of 800 mg/m^2, corresponding to an estimated bioavailability of 40% (see Fig. 2.6). A fixed subcutaneous dose of 1400 mg was subsequently estimated to result in a C_{trough} noninferior to 375 mg/m^2 intravenous rituximab (Bittner et al., 2014). To confirm C_{trough} noninferiority of the selected subcutaneous fixed dose, 154 patients with FL on maintenance therapy were enrolled in the dose-confirmatory stage 2 of the study. Patients were randomized to receive rituximab either subcutaneously at a dose of 1400 mg or intravenously at the approved dose of 375 mg/m^2. Results demonstrated noninferiority in C_{trough} with C_{trough} ratios of subcutaneous/intravenous rituximab of 1.24 (90% CI, 1.02−1.51) and 1.12 (90% CI, 0.86−1.45), respectively (Salar et al., 2014).

Once the final subcutaneous dose had been confirmed and following completion of at least 1 year of intravenous maintenance therapy, patients were allowed to complete treatment on their preferred dosing regimen, an option that was accepted by approximately 40% of patients.

The succeeding international phase 3 study (Sabrina; Davies, Berge, et al., 2017; Davies, Merli, et al., 2017) was designed as an open label, 2-stage trial to compare the pharmacokinetics, efficacy, and safety of subcutaneous versus intravenous rituximab in previously untreated patients with FL. In cycle 1, all 127 patients enrolled in stage 1 (see Fig. 2.7) received an intravenous rituximab

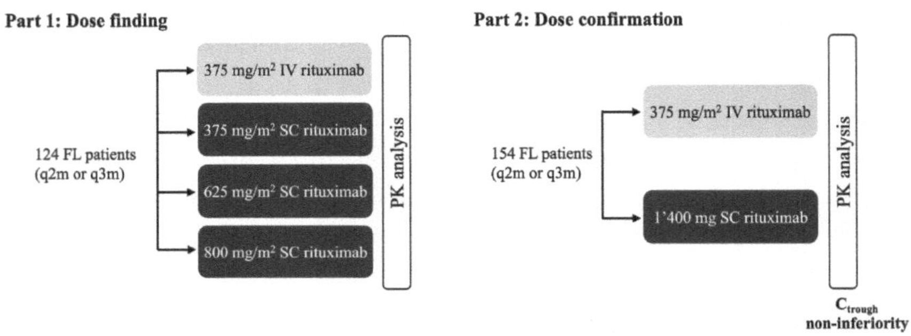

Figure 2.6 Subcutaneous rituximab—dose finding/dose confirmation study. *C_{trough}*, serum trough concentrations; *FL*, follicular lymphoma; *IV*, intravenous; *PK*, pharmacokinetics; *q2m*, every 2 months; *q3m*, every 3 months; *SC*, subcutaneous.

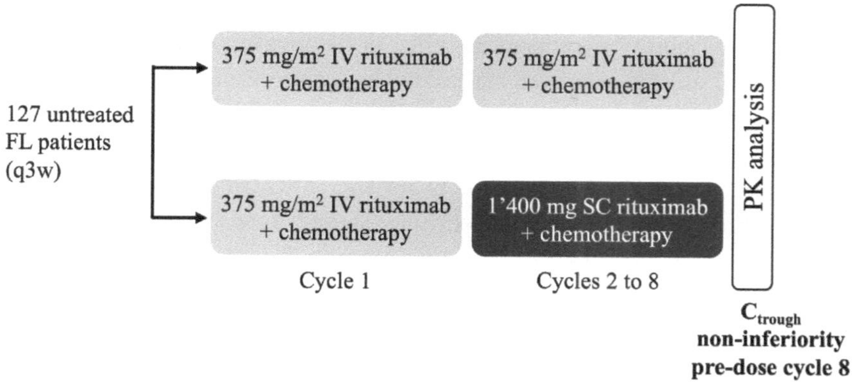

Figure 2.7 Subcutaneous rituximab—phase 3 noninferiority study (stage 1). C_{trough}, serum trough concentrations; *FL*, follicular lymphoma; *IV*, intravenous; *PK*, pharmacokinetics; *q3w*, every 3 weeks; *SC*, subcutaneous.

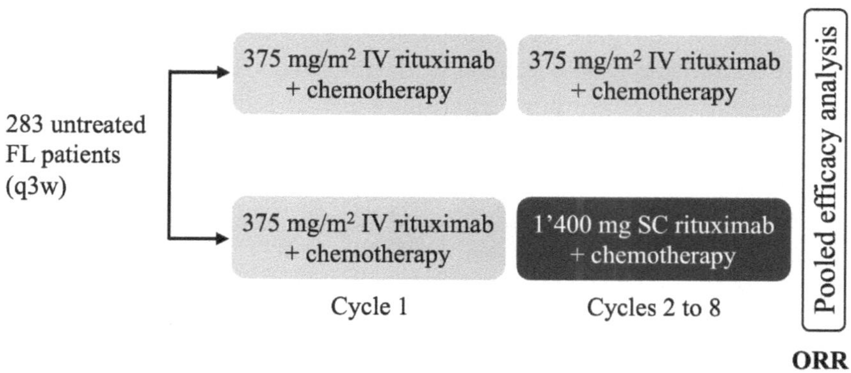

Figure 2.8 Subcutaneous rituximab—phase 3 noninferiority study (stage 2). *FL*, follicular lymphoma; *IV*, intravenous; *ORR*, overall response rate; *q3w*, every 3 weeks; *SC*, subcutaneous.

induction dose of 375 mg/m^2 followed by either subcutaneous doses of 1400 mg or intravenous doses of 375 mg/m^2 q3w for cycles 2–8. To minimize the effect of target-specific elimination of rituximab due to higher tumor load in the early treatment cycles, the noninferiority analysis for C_{trough} was performed in cycle 7. The study met its primary endpoint by demonstrating a C_{trough} ratio (subcutaneous/intravenous) of 1.62 (90% CI, 1.36–1.94). To generate additional clinical evidence including response rates at the end of induction treatment (cycle 8) another 283 patients were randomized in stage 2 (see Fig. 2.8) to receive the same doses and dosing schedules as in stage 1. The preplanned pooled analysis confirmed noninferiority between the tested subcutaneous and intravenous rituximab regimens demonstrating ORR of 84.4% and 84.9%, respectively (Davies, Merli, et al., 2017). Overall, the safety profile of the subcutaneous formulation was comparable to that of the intravenous formulation with the exception of local cutaneous reactions. The incidence of treatment-induced/enhanced antirituximab antibodies was 1.9% and 2% for the intravenous and subcutaneous groups (Electronic Medicines Compendium, 2021).

To obtain supporting clinical data for the subcutaneous formulation in patients with CLL, an indication with a different dose and dosing regimen compared to FL, a third 2-stage clinical study (Sawyer; Assouline et al., 2015; Davies, Berge, et al., 2017) was conducted (see Fig. 2.9). The trial was designed to select and confirm the subcutaneous rituximab dose resulting in noninferior C_{trough} compared to 500 mg/m^2 intravenous rituximab administered q4w. In stage 1 (dose finding), 64 patients initially received 5 cycles of intravenous rituximab starting with a dose of 375 mg/m^2 followed by 4 cycles at a dose of 500 mg/m^2 before receiving a single subcutaneous fixed dose of either 1400, 1600, or 1870 mg. The subcutaneous dose predicted to result in noninferior C_{trough} was 1600 mg. In stage 2 (dose confirmation), 176 patients with untreated CLL were randomized to either subcutaneous or intravenous rituximab at 1600 mg or 500 mg/m^2,

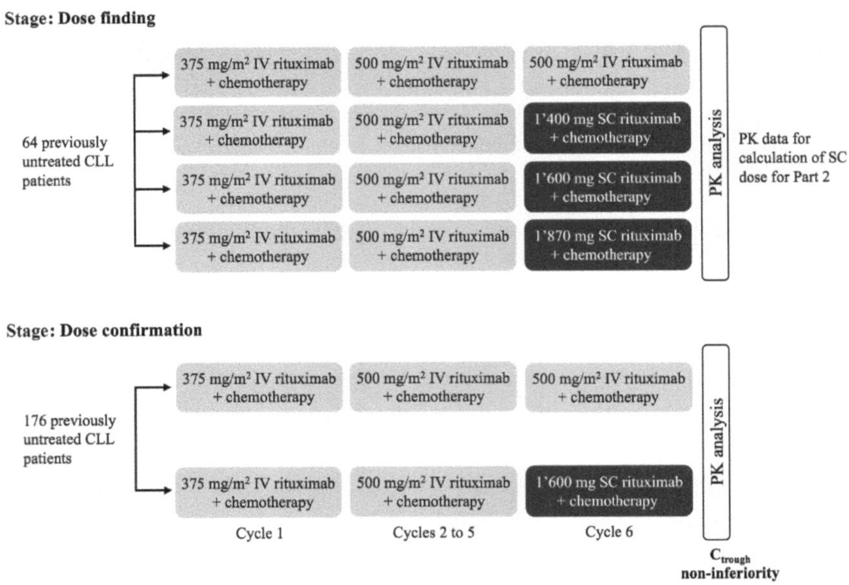

Figure 2.9 Subcutaneous rituximab—dose finding/dose confirmation study in chronic lymphocytic leukemia patients. C_{trough}, serum trough concentrations; *CLL*, chronic lymphocytic leukemia; *IV*, intravenous; *PK*, pharmacokinetics; *q4w*, every 4 weeks; *SC*, subcutaneous.

respectively. Pharmacokinetic noninferiority in rituximab C_{trough} concentrations was assessed at cycle 5 demonstrating noninferiority with an adjusted geometric mean ratio (subcutaneous/intravenous) of 1.53 (90% CI, 1.27–1.85) (Assouline et al., 2016). The demonstrated pharmacokinetic noninferiority complemented with safety and efficacy data supported the EMA approval of subcutaneous rituximab in CLL in 2016 (MabThera; EMA, 2016). Notably, as assessed in a preference questionnaire, more than 90% of patients and nurses preferred subcutaneous administration to standard intravenous infusion (Assouline et al., 2015).

Summary and future perspectives

The development concept for the subcutaneous dosing alternative for trastuzumab in HER2-positive BC formed the basis for the pharmacokinetic-based clinical bridging approach from existing intravenous to subcutaneous administration. The same principles were subsequently successfully applied to rituximab in NHL and CLL indications. Molecule- and indication-specific aspects, such as the dosing frequency or the tolerability profile, were taken into consideration and embedded into the overall approach. The applicability of the pharmacokinetic-based bridging approach is now broadly accepted and was applied to the subcutaneous development programs for daratumumab and the fixed-dose combination of pertuzumab and trastuzumab (Mateos et al., 2020; Tan et al., 2021).

Especially in the context of enabling a flexible care setting that includes the option for home- and patient self-administration, allowing for subcutaneous injections both into the thigh and abdomen is of high importance to ensure maximum possible compliance with the dosing regimen. Today, however, the high-volume subcutaneous dosing alternatives for mAbs in oncology are approved for administration into a single body site only. This is a result of restricting subcutaneous dosing to the use of only one injection area throughout the registrational clinical trials. Due to injection site-dependent absorption through different lymphatic vessels and nodes within the subcutaneous tissue, pharmacokinetic differences can be expected. For future developments, it is recommended that clinical studies should account for the resulting pharmacokinetic variability early on by, for example, giving trial participants the choice to be injected into their preferred site. This way, clinical data for different injection sites are being generated from the beginning of the development program, permitting a broader label as per patient preference.

As it is the case in disease areas, such as rheumatoid arthritis or multiple sclerosis (Bittner & Schmidt, 2021), in the future, also for biological treatments in oncology, clinical developments will start using the subcutaneous formulation from the very beginning. This approach was hindered previously by the challenges associated with the subcutaneous application of high-dose biologicals that have to a great extent been overcome with the introduction of high-concentration formulations, subcutaneous dispersion aids, and the development of high-volume wearable patch injectors (refer to Chapter 1, Enabling customer-centric and sustainable drug delivery—formulation and device lifecycle management of biotherapeutics). In any case, manufacturers need to consider still offering an intravenous dosing form, as this may be warranted in light of intravenously dosed combination partners or due to some patients still preferring intravenous over subcutaneous administration (Rummel et al., 2017). In case only a subcutaneous dosing option is brought forward into pivotal clinical trials of a novel biotherapeutic, the development program is not expected to deviate from what is currently to be shown for intravenous formulations, as the key focus will be on demonstration of efficacy and safety of the new biological molecule. In case both a subcutaneous and intravenous formulation will be developed, phase 3 clinical studies could, for example, include three arms comparing efficacy for both routes of administration to the established standard of care in a given indication. Alternatively, different routes could be offered for different indications for the same biotherapeutic drug, that is, based on the dosing regimen of expected combination partners and preferences.

While this chapter outlines the nonclinical and clinical development studies supporting a science-based bridging approach from an established intravenous formulation to a novel subcutaneous formulation for biologics, it is anticipated that the underlying basic concepts could in principle also serve as precedent for bridging to other alternative administration routes (Philippart et al., 2015).

References

Agbogbo, F. K., Ecker, D. M., Farrand, A., Han, K., Khoury, A., Martin, A., McCool, J., Rasche, U., Rau, T. D., Schmidt, D., Sha, M., & Treuheit, N. (2019). Current perspectives on biosimilars. *Journal of Industrial Microbiology and Biotechnology*, *46*(9–10), 1297–1311. Available from https://doi.org/10.1007/s10295-019-02216-z.

Ahmann, A., Szeinbach, S. L., Gill, J., Traylor, L., & Garg, S. K. (2014). Comparing patient preferences and healthcare provider recommendations with the pen versus vial-and-syringe insulin delivery in patients with type 2 diabetes. *Diabetes Technology & Therapeutics, 16*(2), 76–83. Available from https://doi.org/10.1089/dia.2013.0172.

Altreuter, D. H., Kirtane, A. R., Grant, T., Kruger, C., Traverso, G., & Bellinger, A. M. (2018). Changing the pill: Developments toward the promise of an ultra-long-acting gastroretentive dosage form. *Expert Opinion on Drug Delivery, 15*(12), 1189–1198. Available from https://doi.org/10.1080/17425247.2018.1544615.

Assouline, S., Buccheri, V., Delmer, A., Gaidano, G., McIntyre, C., Brewster, M., Catalani, O., Hourcade-Potelleret, F., Sayyed, P., & Badoux, X. (2015). Pharmacokinetics and safety of subcutaneous rituximab plus fludarabine and cyclophosphamide for patients with chronic lymphocytic leukaemia. *British Journal of Clinical Pharmacology, 80*(5), 1001–1009. Available from https://doi.org/10.1111/bcp.12662.

Assouline, S., Buccheri, V., Delmer, A., Gaidano, G., Trneny, M., Berthillon, N., Brewster, M., Catalani, O., Li, S., McIntyre, C., Sayyed, P., & Badoux, X. (2016). Pharmacokinetics, safety, and efficacy of subcutaneous versus intravenous rituximab plus chemotherapy as treatment for chronic lymphocytic leukaemia (SAWYER): A phase 1b, open-label, randomised controlled non-inferiority trial. *The Lancet Haematology, 3*(3), e128–e138. Available from https://doi.org/10.1016/S2352-3026(16)00004-1.

Bai, S., Jorga, K., Xin, Y., Jin, D., Zheng, Y., Damico-Beyer, L. A., Gupta, M., Tang, M., Allison, D. E., Lu, D., Zhang, Y., Joshi, A., & Dresser, M. J. (2012). A guide to rational dosing of monoclonal antibodies. *Clinical Pharmacokinetics, 51*(2), 119–135. Available from https://doi.org/10.2165/11596370-000000000-00000.

Baselga, J., Carbonell, X., Castañeda-Soto, N. J., Clemens, M., Green, M., Harvey, V., Morales, S., Barton, C., & Ghahramani, P. (2005). Phase II study of efficacy, safety, and pharmacokinetics of trastuzumab monotherapy administered on a 3-weekly schedule. *Journal of Clinical Oncology, 23*(10), 2162–2171. Available from https://doi.org/10.1200/JCO.2005.01.014.

Berinstein, N. L., Grillo-López, A. J., White, C. A., Bence-Bruckler, I., Maloney, D., Czuczman, M., Green, D., Rosenberg, J., McLaughlin, P., & Shen, D. (1998). Association of serum rituximab (IDEC-C2B8) concentration and anti-tumor response in the treatment of recurrent low-grade or follicular non-Hodgkin's lymphoma. *Annals of Oncology, 9*(9), 995–1001. Available from https://doi.org/10.1023/A:1008416911099.

Bittner, B., Richter, W. F., Hourcade-Potelleret, F., Herting, F., & Schmidt, J. (2014). Non-clinical pharmacokinetic/pharmacodynamic and early clinical studies supporting development of a novel subcutaneous formulation for the monoclonal antibody rituximab. *Drug Research, 64*(11), 569–575. Available from https://doi.org/10.1055/s-0033-1363993.

Bittner, B., Richter, W. F., Hourcade-Potelleret, F., McIntyre, C., Herting, F., Zepeda, M. L., & Schmidt, J. (2012). Development of a subcutaneous formulation for trastuzumab-nonclinical and clinical bridging approach to the approved intravenous dosing regimen. *Arzneimittelforschung, 62*(9), 401–409. Available from https://doi.org/10.1055/s-0032-1321831.

Bittner, B., Richter, W. F., & Schmidt, J. (2018). Subcutaneous administration of biotherapeutics: An overview of current challenges and opportunities. *BioDrugs, 32*(5), 425–440. Available from https://doi.org/10.1007/s40259-018-0295-0.

Bittner, B., & Schmidt, J. (2021). Subcutaneous drug delivery devices—Enablers of a flexible care setting. In E. Chappel (Ed.), *Drug delivery devices and therapeutic systems* (pp. 159–179). Elsevier BV. Available from https://doi.org/10.1016/b978-0-12-819838-4.00021-3.

Blackwood, C., Sanga, P., Nuamah, I., Keenan, A., Singh, A., Mathews, M., & Gopal, S. (2020). Patients' preference for long-acting injectable versus oral antipsychotics in schizophrenia: Results from the patient-reported medication preference questionnaire. *Patient Preference and Adherence, 14*, 1093–1102. Available from https://doi.org/10.2147/PPA.S251812.

Boccara, F., Dent, R., Ruilope, L., & Valensi, P. (2017). Practical considerations for the use of subcutaneous treatment in the management of dyslipidaemia. *Advances in Therapy*, *34* (8), 1876–1896. Available from https://doi.org/10.1007/s12325-017-0586-8.

Bookbinder, L. H., Hofer, A., Haller, M. F., Zepeda, M. L., Keller, G. A., Lim, J. E., Edgington, T. S., Shepard, H. M., Patton, J. S., & Frost, G. I. (2006). A recombinant human enzyme for enhanced interstitial transport of therapeutics. *Journal of Controlled Release*, *114*(2), 230–241. Available from https://doi.org/10.1016/j.jconrel.2006.05.027.

Brennan, F. R., Morton, L. D., Spindeldreher, S., Kiessling, A., Allenspach, R., Hey, A., Muller, P. Y., Frings, W., & Sims, J. (2010). Safety and immunotoxicity assessment of immunomodulatory monoclonal antibodies. *MAbs*, *2*(3), 233–255. Available from https://doi.org/10.4161/mabs.2.3.11782.

Burcombe, R., Chan, S., Simcock, R., Samanta, K., Percival, F., & Barrett-Lee, P. (2013). Subcutaneous trastuzumab (Herceptin®): A UK time and motion study in comparison with intravenous formulation for the treatment of patients with HER2-positive early breast cancer. *Advances in Breast Cancer Research*, *2*, 133–140. Available from https://doi.org/10.4236/abcr.2013.24022.

Burmester, G. R., Rubbert-Roth, A., Cantagrel, A., Hall, S., Leszczynski, P., Feldman, D., Rangaraj, M. J., Roane, G., Ludivico, C., Lu, P., Rowell, L., Bao, M., & Mysler, E. F. (2014). A randomised, double-blind, parallel-group study of the safety and efficacy of subcutaneous tocilizumab versus intravenous tocilizumab in combination with traditional disease-modifying antirheumatic drugs in patients with moderate to severe rheumatoid arthritis (SUMMACTA study). *Annals of the Rheumatic Diseases*, *73*(1), 69–74. Available from https://doi.org/10.1136/annrheumdis-2013-203523.

Connor, R. J., Taverna, D. M., Thrall, K., LaBarre, M. J., & Kang, D. W. (2020). Use of computed tomography to assess subcutaneous drug dispersion with recombinant human hyaluronidase PH20 in a swine model. *Journal of Pharmacological and Toxicological Methods*, 106. Available from https://doi.org/10.1016/j.vascn.2020.106936.

Curigliano, G., O'Connor, D. P., Rosenberg, J. A., & Jacobs, I. (2016). Biosimilars: Extrapolation for oncology. *Critical Reviews in Oncology/Hematology*, *104*, 131–137. Available from https://doi.org/10.1016/j.critrevonc.2016.06.002.

Davies, A., Berge, C., Boehnke, A., Dadabhoy, A., Lugtenburg, P., Rule, S., Rummel, M., McIntyre, C., Smith, R., & Badoux, X. (2017). Subcutaneous rituximab for the treatment of B-cell hematologic malignancies: A review of the scientific rationale and clinical development. *Advances in Therapy*, *34*(10), 2210–2231. Available from https://doi.org/10.1007/s12325-017-0610-z.

Davies, A., Merli, F., Mihaljević, B., Mercadal, S., Siritanaratkul, N., Solal-Céligny, P., Boehnke, A., Berge, C., Genevray, M., Zharkov, A., Dixon, M., Brewster, M., Barrett, M., & MacDonald, D. (2017). Efficacy and safety of subcutaneous rituximab versus intravenous rituximab for first-line treatment of follicular lymphoma (SABRINA): A randomised, open-label, phase 3 trial. *The Lancet Haematology*, *4*(6), e272–e282. Available from https://doi.org/10.1016/S2352-3026(17)30078-9.

Deng, J., Zhu, X., Chen, Z., Fan, C. H., Kwan, H. S., Wong, C. H., Shek, K. Y., Zuo, Z., & Lam, T. N. (2017). A review of food-drug interactions on oral drug absorption. *Drugs*, 77(17), 1833–1855. Available from https://doi.org/10.1007/s40265-017-0832-z.

Egorin, M. J. (2003). Horseshoes, hand grenades, and body-surface area-based dosing: Aiming for a target. *Journal of Clinical Oncology*, *21*(2), 182–183. Available from https://doi.org/10.1200/JCO.2003.10.084.

Electronic Medicines Compendium. (2021). <https://www.medicines.org.uk/emc/>.

EMA. (2008). *Guideline on immunogenicity assessment of biotechnology-derived therapeutic proteins, EMEA/CHMP/BMWP/14327/2006* <https://www.ema.europa.eu/en/documents/scientific-guideline/draft-guideline-immunogenicity-assessment-biotechnology-derived-therapeutic-proteins-revision-1_en.pdf>.

EMA. *Guideline on similar biological medicinal products containing monoclonal antibodies – Non-clinical and clinical issues, EMA/CHMP/BMWP/403543/2010.* (2012). <https://www.ema.europa.eu/en/documents/scientific-guideline/guideline-similar-biological-medicinal-products-containing-monoclonal-antibodies-non-clinical_en.pdf>.
EMA. *Herceptin®, CHMP assessment report, procedure no. EMEA/H/C/000278 27.* (2013). <https://www.ema.europa.eu/en/documents/variation-report/herceptin-h-c-278-x-0060-epar-assessment-report-extension_en.pdf>.
EMA. *Guideline on similar biological medicinal products, CHMP/437/04 Rev 1.* (2014a). <https://www.ema.europa.eu/en/documents/scientific-guideline/guideline-similar-biological-medicinal-products-rev1_en.pdf>.
EMA. *Guideline on the pharmacokinetic and clinical evaluation of modified release dosage forms, EMA/CHMP/EWP/280/96 Rev1.* (2014b). <https://www.ema.europa.eu/en/documents/scientific-guideline/guideline-pharmacokinetic-clinical-evaluation-modified-release-dosage-forms_en.pdf>.
EMA. *MabThera®, CHMP assessment report, procedure no. EMEA/H/C/000165/X/0083.* (2014c). <https://www.ema.europa.eu/en/documents/variation-report/mabthera-h-c-165-x-83-epar-assessment-report-extension_en.pdf>.
EMA. *MabThera®, CHMP assessment report, procedure no. EMEA/H/C/000165/X/0101/G.* (2016). <https://www.ema.europa.eu/en/documents/variation-report/mabthera-h-c-165-x-101-g-epar-assessment-report-variation_en.pdf>.
EMA. *Guideline on clinical development of fixed combination medicinal products, EMA/CHMP/158268/2017.* (2017a). <https://www.ema.europa.eu/en/documents/scientific-guideline/guideline-clinical-development-fixed-combination-medicinal-products-revision-2_en.pdf>.
EMA. *Reflection paper on the dissolution specification for generic solid oral immediate release products with systemic action, EMA/CHMP/CVMP/QWP/336031/2017, Committee for Medicinal Products for Human use (CHMP).* (2017b). <https://www.ema.europa.eu/en/documents/scientific-guideline/reflection-paper-dissolution-specification-generic-solid-oral-immediate-release-products-systemic_en.pdf>.
EMA. *ICH M9 guideline on biopharmaceutics classification system-based biowaivers, EMA/CHMP/ICH/493213/2018, Committee for Medicinal Products for Human Use.* (2020). <https://www.ema.europa.eu/en/documents/scientific-guideline/ich-m9-biopharmaceutics-classification-system-based-biowaivers-step-5_en.pdf>.
FDA. *Exposure-response relationships—Study design, data analysis, and regulatory applications, FDA guidance for industry.* (2003). <https://www.fda.gov/media/71277/download>.
FDA. *Fixed dose combinations, co-packaged drug products, and single-entity versions of previously approved antiretrovirals for the treatment of HIV, FDA guidance for industry.* (2006). <https://www.fda.gov/media/72248/download>.
FDA. *Immunogenicity assessment for therapeutic protein products, FDA guidance for industry.* (2014). <https://www.fda.gov/media/85017/download>.
FDA. *Clinical pharmacology data to support a demonstration of biosimilarity to a reference product, FDA guidance for industry.* (2016). <https://www.fda.gov/media/88622/download>.
FDA. *Biological product definitions.* (2017). <https://www.fda.gov/downloads/Drugs/DevelopmentApprovalProcess/HowDrugsareDevelopedandApproved/ApprovalApplications/TherapeuticBiologicApplications/Biosimilars/UCM581282.pdf>.
FDA. *Considerations in demonstrating interchangeability with a reference product, FDA guidance for industry.* (2019). <https://www.fda.gov/media/124907/download>.
FDA Label. (2021). <https://nctr-crs.fda.gov/fdalabel/ui/search>.
Freshwater, T., Kondic, A., Ahamadi, M., Li, C. H., de Greef, R., de Alwis, D., & Stone, J. A. (2017). Evaluation of dosing strategy for pembrolizumab for oncology indications. *Journal for ImmunoTherapy of Cancer, 5*(1). Available from https://doi.org/10.1186/s40425-017-0242-5.

Golay, J., Semenzato, G., Rambaldi, A., Foà, R., Gaidano, G., Gamba, E., Pane, F., Pinto, A., Specchia, G., Zaja, F., & Regazzi, M. (2013). Lessons for the clinic from rituximab pharmacokinetics and pharmacodynamics. *MAbs*, *5*(6), 826–837. Available from https://doi.org/10.4161/mabs.26008.

Golchin, N., Frank, S., Vince, A., Isham, L., & Meropol, S. (2015). Polypharmacy in the elderly. *Journal of Research in Pharmacy Practice*, *4*(2), 85–88. Available from https://doi.org/10.4103/2279-042X.155755.

Hall, E., Zhang, J., Kim, E. J., Hwang, G., Chu, G., Bhatia, S., & Reddy, S. (2020). Economics of alternative dosing strategies for pembrolizumab and nivolumab at a single academic cancer center. *Cancer Medicine*, *9*(6), 2106–2112. Available from https://doi.org/10.1002/cam4.2888.

Hegg, R., Pienkowski, T., Chen, S., Staroslawska, E., Falcon, S., Kovalenko, N., Al-Sakaff, N., Heinzmann, D., Kolaitis, G., & Ismael, G. (2012). 273P—Immunogenicity of trastuzumab intravenous and subcutaneous formulations in the phase III HannaH study. *Annals of Oncology*, *23*, ix103. Available from https://doi.org/10.1016/S0923-7534(20)32835-0.

Hendrikx, J. J. M. A., Haanen, J. B. A. G., Voest, E. E., Schellens, J. H. M., Huitema, A. D. R., & Beijnen, J. H. (2017). Fixed dosing of monoclonal antibodies in oncology. *Oncologist*, *22*(10), 1212–1221. Available from https://doi.org/10.1634/theoncologist.2017-0167.

Hourcade-Potelleret, F., Lemenuel-Diot, A., McIntyre, C., Brewster, M., Lum, B., & Bittner, B. (2014). Use of a population pharmacokinetic approach for the clinical development of a fixed-dose subcutaneous formulation of trastuzumab. *CPT Pharmacometrics Systems Pharmacology*, *3*, e87. Available from https://doi.org/10.1038/psp.2013.63.

Hsu, A., Yao, H. M., Gupta, S., & Modi, N. B. (2015). Comparison of the pharmacokinetics of an oral extended-release capsule formulation of carbidopa-levodopa (IPX066) with immediate-release carbidopa-levodopa (Sinemet®), sustained-release carbidopa-levodopa (Sinemet® CR), and carbidopa-levodopa-entacapone (Stalevo®). *Journal of Clinical Pharmacology*, *55*(9), 995–1003. Available from https://doi.org/10.1002/jcph.514.

Ismael, G., Hegg, R., Muehlbauer, S., Heinzmann, D., Lum, B., Kim, S. B., Pienkowski, T., Lichinitser, M., Semiglazov, V., Melichar, B., & Jackisch, C. (2012). Subcutaneous versus intravenous administration of (neo)adjuvant trastuzumab in patients with HER2-positive, clinical stage I-III breast cancer (HannaH study): A phase 3, open-label, multicentre, randomised trial. *The Lancet Oncology*, *13*(9), 869–878. Available from https://doi.org/10.1016/S1470-2045(12)70329-7.

Jackisch, C., Scappaticci, F. A., Heinzmann, D., Bisordi, F., Schreitmüller, T., Minckwitz, G. V., & Cortés, J. (2015). Neoadjuvant breast cancer treatment as a sensitive setting for trastuzumab biosimilar development and extrapolation. *Future Oncology*, *11*(1), 61–71. Available from https://doi.org/10.2217/fon.14.187.

Jackisch, C., Stroyakovskiy, D., Pivot, X., Ahn, J. S., Melichar, B., Chen, S. C., Meyenberg, C., Al-Sakaff, N., Heinzmann, D., & Hegg, R. (2019). Subcutaneous versus intravenous trastuzumab for patients with ERBB2-positive early breast cancer: Final analysis of the HannaH phase 3 randomized clinical trial. *JAMA Oncology*, *5*(5), e190339. Available from https://doi.org/10.1001/jamaoncol.2019.0339.

Jolles, S., Bernatowska, E., de Gracia, J., Borte, M., Cristea, V., Peter, H. H., Belohradsky, B. H., Wahn, V., Neufang-Hüber, J., Zenker, O., & Grimbacher, B. (2011). Efficacy and safety of Hizentra® in patients with primary immunodeficiency after a dose-equivalent switch from intravenous or subcutaneous replacement therapy. *Clinical Immunology*, *141*(1), 90–102. Available from https://doi.org/10.1016/j.clim.2011.06.002.

Kivitz, A., Olech, E., Borofsky, M., Zazueta, B. M., Navarro-Sarabia, F., Radominski, S. C., Merrill, J. T., Rowell, L., Nasmyth-Miller, C., Bao, M., Wright, S., & Pope, J. E. (2014). Subcutaneous tocilizumab versus placebo in combination with disease-modifying antirheumatic drugs in patients with rheumatoid arthritis. *Arthritis Care and Research*, *66*(11), 1653–1661. Available from https://doi.org/10.1002/acr.22384.

Kolberg, H. C., Colleoni, M., Santi, P., Demetriou, G. S., Segui-Palmer, M. A., Fujiwara, Y., Hurvitz, S. A., & Hanes, V. (2019). Totality of scientific evidence in the development of ABP 980, a biosimilar to trastuzumab. Targeted. *Oncology*, *14*(6), 647–656. Available from https://doi.org/10.1007/s11523-019-00675-z.

Mahmood, I., Tegenge, M. A., & Golding, B. (2020). Considerations for optimizing dosing of immunoglobulins based on pharmacokinetic evidence. *Antibodies*, 24. Available from https://doi.org/10.3390/antib9020024.

Masnoon, N., Shakib, S., Kalisch-Ellett, L., & Caughey, G. E. (2017). What is polypharmacy? A systematic review of definitions. *BMC Geriatrics*, *17*(1), 230. Available from https://doi.org/10.1186/s12877-017-0621-2.

Mateos, M. V., Nahi, H., Legiec, W., Grosicki, S., Vorobyev, V., Spicka, I., Hungria, V., Korenkova, S., Bahlis, N., Flogegard, M., Bladé, J., Moreau, P., Kaiser, M., Iida, S., Laubach, J., Magen, H., Cavo, M., Hulin, C., White, D., & Usmani, S. Z. (2020). Subcutaneous versus intravenous daratumumab in patients with relapsed or refractory multiple myeloma (COLUMBA): A multicentre, open-label, non-inferiority, randomised, phase 3 trial. *The Lancet Haematology*, 7(5), e370–e380. Available from https://doi.org/10.1016/S2352-3026(20)30070-3.

McLennan, D. N., Porter, C., & Charman, S. A. (2005). Subcutaneous drug delivery and the role of the lymphatics. *Drug Discovery Today: Technologies*, *2*(1), 89–96. Available from https://doi.org/10.1016/j.ddtec.2005.05.006.

Monaco, A., Palmer, K., Faber, N. H. R., Kohler, I., Silva, M., Vatland, A., van Griensven, J., Votta, M., Walsh, D., Clay, V., Yazicioglu, M. C., Ducinskiene, D., & Donde, S. (2021). Digital health tools for managing noncommunicable diseases during and after the COVID-19 pandemic: Perspectives of patients and caregivers. *Journal of Medical Internet Research*, *23*(1). Available from https://doi.org/10.2196/25652.

Mott, D. J. (2018). Incorporating quantitative patient preference data into healthcare decision making processes: Is HTA falling behind? *Patient*, *11*(3), 249–252. Available from https://doi.org/10.1007/s40271-018-0305-9.

Ogungbenro, K., Patel, A., Duncombe, R., Nuttall, R., Clark, J., & Lorigan, P. (2018). Dose rationalization of pembrolizumab and nivolumab using pharmacokinetic modeling and simulation and cost analysis. *Clinical Pharmacology and Therapeutics*, *103*(4), 582–590. Available from https://doi.org/10.1002/cpt.875.

Paul, F., & Cartron, G. (2019). Infusion-related reactions to rituximab: Frequency, mechanisms and predictors. *Expert Review of Clinical Immunology*, *15*(4), 383–389. Available from https://doi.org/10.1080/1744666X.2019.1562905.

Philippart, M., Schmidt, J., & Bittner, B. (2015). Oral delivery of therapeutic proteins and peptides: An overview of current technologies and recommendations for bridging from approved intravenous or subcutaneous administration to novel oral regimens. *Drug Research*, *66*(3), 113–120. Available from https://doi.org/10.1055/s-0035-1559654.

Reilly, J. J., & Workman, P. (1993). Normalisation of anti-cancer drug dosage using body weight and surface area: Is it worthwhile? *Cancer Chemotherapy and Pharmacology*, *32*, 411–418. Available from https://doi.org/10.1007/bf00685883.

Richter, W. F., & Jacobsen, B. (2014). Subcutaneous absorption of biotherapeutics: Knowns and unknowns. *Drug Metabolism and Disposition*, *42*(11), 1881–1889. Available from https://doi.org/10.1124/dmd.114.059238.

Rule, S., Collins, G. P., & Samanta, K. (2014). Subcutaneous versus intravenous rituximab in patients with non-Hodgkin lymphoma: A time and motion study in the United

Kingdom. *Journal of Medical Economics*, *17*(7), 459–468. Available from https://doi.org/10.3111/13696998.2014.914033.

Rummel, M., Kim, T. M., Aversa, F., Brugger, W., Capochiani, E., Plenteda, C., Re, F., Trask, P., Osborne, S., Smith, R., & Grigg, A. (2017). Preference for subcutaneous or intravenous administration of rituximab among patients with untreated CD20 + diffuse large B-cell lymphoma or follicular lymphoma: Results from a prospective, randomized, open-label, crossover study (PrefMab). *Annals of Oncology*, *28*(4), 836–842. Available from https://doi.org/10.1093/annonc/mdw685.

Ryman, J. T., & Meibohm, B. (2017). Pharmacokinetics of monoclonal antibodies. *CPT Pharmacometrics and Systems Pharmacology*, *6*, 576–588. Available from https://doi.org/10.1002/psp4.12224.

Salar, A., Avivi, I., Bittner, B., Bouabdallah, R., Brewster, M., Catalani, O., Follows, G., Haynes, A., Hourcade-Potelleret, F., Janikova, A., Larouche, J. F., McIntyre, C., Pedersen, M., Pereira, J., Sayyed, P., Shpilberg, O., & Tumyan, G. (2014). Comparison of subcutaneous versus intravenous administration of rituximab as maintenance treatment for follicular lymphoma: Results from a two-stage, phase Ib study. *Journal of Clinical Oncology*, *32*(17), 1782–1791. Available from https://doi.org/10.1200/JCO.2013.52.2631.

Schernthaner, G. (2010). Fixed-dose combination therapies in the management of hyperglycaemia in type 2 diabetes: An opportunity to improve adherence and patient care. *Diabetic Medicine*, *27*(7), 739–743. Available from https://doi.org/10.1111/j.1464-5491.2010.03000.x.

Shpilberg, O., & Jackisch, C. (2013). Subcutaneous administration of rituximab (MabThera®) and trastuzumab (Herceptin®) using hyaluronidase. *British Journal of Cancer*, *109*(6), 1556–1561. Available from https://doi.org/10.1038/bjc.2013.371.

Srivastava, K., Arora, Kataria, Cappelleri, J. C., Sadosky., & Peterson, A. M. (2013). Impact of reducing dosing frequency on adherence to oral therapies: A literature review and *meta*-analysis. *Patient Prefer Adherence*, 7, 419–434. Available from https://doi.org/10.2147/PPA.S44646.

Stoner, K. L., Harder, H., Fallowfield, L. J., & Jenkins, V. A. (2015). Intravenous versus subcutaneous drug administration. Which do patients prefer? A systematic review. *Patient*, *8*(2), 145–153. Available from https://doi.org/10.1007/s40271-014-0075-y.

SUPAC-IR. *Immediate-release solid oral dosage forms: Scale-up and post-approval changes: Chemistry, manufacturing and controls*. (1995). <https://www.fda.gov/media/70949/download>.

SUPAC-MR. *Modified release solid oral dosage forms: Scale-up and post-approval changes: Chemistry, manufacturing, and controls*. (1997). <https://www.fda.gov/media/70956/download>.

Tan, A. R., Im, S. A., Mattar, A., Colomer, R., Stroyakovskii, D., Nowecki, Z., De Laurentiis, M., Pierga, J. Y., Jung, K. H., Schem, C., Hogea, A., Badovinac Crnjevic, T., Heeson, S., Shivhare, M., Kirschbrown, W. P., Restuccia, E., & Jackisch, C. FeDeriCa Study Group. (2021). Fixed-dose combination of pertuzumab and trastuzumab for subcutaneous injection plus chemotherapy in HER2-positive early breast cancer (FeDeriCa): A randomised, open-label, multicentre, non-inferiority, phase 3 study. *The Lancet Oncology*, *22*, 85–97. Available from https://doi.org/10.1016/S1470-2045(20)30536-2.

Ternant, D., Monjanel, H., Venel, Y., Prunier-Aesch, C., Arbion, F., Colombat, P., Paintaud, G., & Gyan, E. (2019). Nonlinear pharmacokinetics of rituximab in non-Hodgkin lymphomas: A pilot study. *British Journal of Clinical Pharmacology*, *85*(9), 2002–2010. Available from https://doi.org/10.1111/bcp.13991.

Ubaldo, J. M., Fuhr, R., Forte, P., Barker, P., Milton, J. A., Aurivillius, M., Yan, L., & Roskos, L. (2021). Comparison of autoinjector with accessorized prefilled syringe for benralizumab pharmacokinetic exposure: AMES trial results. *Journal of Asthma*, *58*(1), 93–101. Available from https://doi.org/10.1080/02770903.2019.1663428.

Usach, I., Martinez, R., Festini, T., & Peris, J. E. (2019). Subcutaneous injection of drugs: Literature review of factors influencing pain sensation at the injection site. *Advances in Therapy*, *36*(11), 2986–2996. Available from https://doi.org/10.1007/s12325-019-01101-6.

van Overbeeke, E., Whichello, C., Janssens, R., Veldwijk, J., Cleemput, I., Simoens, S., Juhaeri, J., Levitan, B., Kübler, J., de Bekker-Grob, E., & Huys, I. (2019). Factors and situations influencing the value of patient preference studies along the medical product lifecycle: A literature review. *Drug Discovery Today*, *24*(1), 57–68. Available from https://doi.org/10.1016/j.drudis.2018.09.015.

Varkhede, N., & Forrest, M. L. (2018). Understanding the monoclonal antibody disposition after subcutaneous administration using a minimal physiologically based pharmacokinetic model. *Journal of Pharmacy and Pharmaceutical Sciences*, *21*(1S), 130–148. Available from https://doi.org/10.18433/jpps30028.

Viola, M., Sequeira, J., Seiça, R., Veiga, F., Serra, J., Santos, A. C., & Ribeiro, A. J. (2018). Subcutaneous delivery of monoclonal antibodies: How do we get there? *Journal of Controlled Release*, *286*, 301–314. Available from https://doi.org/10.1016/j.jconrel.2018.08.001.

Vogel, C. L., Cobleigh, M. A., Tripathy, D., Gutheil, J. C., Harris, L. N., Fehrenbacher, L., Slamon, D. J., Murphy, M., Novotny, W. F., Burchmore, M., Shak, S., Stewart, S. J., & Press, M. (2002). Efficacy and safety of trastuzumab as a single agent in first-line treatment of HER2-overexpressing metastatic breast cancer. *Journal of Clinical Oncology*, *20*(3), 719–726. Available from https://doi.org/10.1200/JCO.20.3.719.

Wai, A. O., Frighetto, L., Marra, C. A., Chan, E., & Jewesson, P. J. (2000). Cost analysis of an adult outpatient parenteral antibiotic therapy (OPAT) programme: A Canadian teaching hospital and ministry of health perspective. *PharmacoEconomics*, *18*(5), 451–457. Available from https://doi.org/10.2165/00019053-200018050-00004.

Wang, D. D., Zhang, S., Zhao, H., Men, A. Y., & Parivar, K. (2009). Fixed dosing versus body size-based dosing of monoclonal antibodies in adult clinical trials. *Journal of Clinical Pharmacology*, *49*(9), 1012–1024. Available from https://doi.org/10.1177/0091270009337512.

Wheless, J. W., & Phelps, S. J. (2018). A clinician's guide to oral extended-release drug delivery systems in epilepsy. *Journal of Pediatric Pharmacology and Therapeutics*, *23*(4), 277–292. Available from https://doi.org/10.5863/1551-6776-23.4.277.

Wynne, C., Harvey, V., Schwabe, C., Waaka, D., McIntyre, C., & Bittner, B. (2013). Comparison of subcutaneous and intravenous administration of trastuzumab: A phase I/Ib trial in healthy male volunteers and patients with HER2-positive breast cancer. *Journal of Clinical Pharmacology*, *53*(2), 192–201. Available from https://doi.org/10.1177/0091270012436560.

Yin, A., Li, J., Hurst, D., & Visich, J. (2010). Population pharmacokinetics (PK) and association of PK and clinical outcomes of rituximab in patients with non-Hodgkin's lymphoma. *Journal of Clinical Oncology*, *28*(15). Available from https://doi.org/10.1200/jco.2010.28.15_suppl.e13108, e13108.

Zhao, X., Suryawanshi, S., Hruska, M., Feng, Y., Wang, X., Shen, J., Vezina, H. E., McHenry, M. B., Waxman, I. M., Achanta, A., Bello, A., Roy, A., & Agrawal, S. (2017). Assessment of nivolumab benefit-risk profile of a 240-mg flat dose relative to a 3-mg/kg dosing regimen in patients with advanced tumors. *Annals of Oncology*, *28*(8), 2002–2008. Available from https://doi.org/10.1093/annonc/mdx235.

Zharkov, A., Barton, B., Heinzmann, D., Bakalos, G., & Schreitmüller, T. (2019). Development pathways for subcutaneous formulations of biologics versus biosimilar development. *Expert Review of Precision Medicine and Drug Development*, 1–8. Available from https://doi.org/10.1080/23808993.2019.1585806.

CHAPTER 3

Clinical development of automated subcutaneous injection devices—established pathways and novel concepts

Introduction

Chapter 2 discussed the development of subcutaneous dosing alternatives to intravenous infusions for monoclonal antibodies (mAbs) aiming at simplifying parenteral drug administration. As described, subcutaneous delivery with fixed dosing regimens is broadly preferred by patients and healthcare providers and has the potential to reduce healthcare costs and resources associated with parenteral drug administration. At initial launch of a novel biological therapy, the product is typically available with a prefilled syringe (PFS) for dosing volumes of up to approximately 2 mL or with a vial, especially when the dosing volume exceeds 2 mL or more. The higher volume formulation from the vial presentation is injected manually by means of a handheld syringe (HHS) or an infusion set (IS).

To further increase patient experience with the dosing regimen and to enable a flexible care setting that includes home- and patient self-administration, the development of automated injection devices, such as autoinjectors (AI) or higher volume on-body delivery systems (OBDS), is warranted. Administering a new mAb in an established device platform is expected to create a broader patient and healthcare provider familiarity with the device, and available learnings on device functionality would inform regular device upgrades and the design of future platforms. Notably, while subcutaneous AI devices are established platforms, the majority of OBDS devices for high-volume delivery is still in development (Bittner & Schmidt, 2021).

Unless an available device platform can be leveraged for new molecules, the biologic–device combination is frequently not yet available for inclusion into the pivotal clinical trials and thus respective clinical device development is conducted as a subsequent lifecycle management activity. In such cases,

Formulation and Device Lifecycle Management of Biotherapeutics
DOI: https://doi.org/10.1016/B978-0-12-823741-0.00009-9
© 2022 Elsevier Inc.
All rights reserved.

the prevalent clinical development approach is to demonstrate appropriate handling of the device by patients, lay caregivers, or healthcare providers in the intended use environment. In addition, pharmacokinetic comparability between the mode of administration used in the pivotal study (PFS or HHS/IS) and the administration via the automated injection device (AI or OBDS) needs to be demonstrated.

This chapter describes common bridging approaches to automated device technologies that have not been an integral part of the pivotal clinical trials for a novel biological molecule. Moreover, a concept on how to leverage existing knowledge to further streamline the underlying clinical development programs for device platforms for mAbs is proposed and discussed based on practical considerations. Technical device development aspects are off-scope.

Usability and pharmacokinetic comparability studies for biologic–device combination products

The following section describes the requirements for biologic–device combinations in which the biological agent and the device form a single integrated product. In the subcutaneous space, this includes PFS or AI and pen devices (both referred to as AI in the following) for dosing volumes of up to approximately 2 mL, and OBDS for higher injection volumes. Common subcutaneous injection device types have been summarized previously (Bittner & Schmidt, 2021).

Usability studies

To support approval of a novel subcutaneous biologic–device combination, manufacturers are required to demonstrate that the product can be used safely by the patient or healthcare provider to deliver the medication according to the intended label instructions (CHMP, 2019; FDA, 2013, 2016, 2019). Existing regulatory guidance provides a sound basis for design, development, and usability requirements. However, instructions also leave room for defining a robust qualification program generating evidence that the product delivers the accurate dosing volume and can be operated in the intended use environment.

Development of an injection device is an iterative process and benefits from so-called formative human factors studies (FDA, 2016; MHRA, 2021) to guide design changes during prototype development. These trials assess whether a patient could safely and effectively use the prototype device and also evaluate the effectiveness of the draft instructions for use and labeling guide (CHMP, 2019; Newswanger et al., 2019). Study participants can include patients or a population that represents the target patient population with respect to age, gender, and expected self-administration capabilities (Lange & Nemeth, 2018), lay caregivers, and healthcare providers (Domańska et al., 2018).

Following completion of the device design verification with focus on product conformance to specifications, product functionality and process capability (Maropoulos & Ceglarek, 2010), summative human factors validation study(ies) to identify and mitigate potential use errors (Rojas et al., 2019) are performed with the final to be marketed biologic–device combination. Studies assess device design, ease of use, intuitiveness, and risks of the device and determine whether the device can be correctly, safely, and effectively used by the intended user population. Moreover, the device labeling and instructions for use are assessed with respect to support users in mitigating high risks and use contexts (CHMP, 2019).

During formative and summative human factors trials, injections are typically done into an injection pad to simulate the subcutaneous administration procedure (Brand-Schieber et al., 2016). While in many cases such noninterventional simulated use studies are sufficient to support the registration of a biologic–device combination, there can be cases where drug administration in a real-world environment could be required to provide supporting evidence on the usability of the product. Such "real use" or "actual use" studies would involve a real subcutaneous injection and could be conducted as a dedicated stand-alone study (Fettner et al., 2019) or combined with clinical trials (Clark & Forkey, 2018). Data generated in an actual use study can provide insights into any discomfort at the injection site experienced by the patient and thus help further mitigating the frequency of this risk (Travanty et al., 2018).

Pharmacokinetic comparability studies

In addition to demonstrating safe and effective use of a novel automated subcutaneous injection device, in most cases, studies that assess a possible impact of the device on the pharmacokinetic profile of a biological molecule are part of the clinical bridging pathway. Particularly, if the device was not yet integrated into the pivotal trial supporting registration of the subcutaneous formulation, manufacturers need to demonstrate that the use of different device modalities does not impair the efficacy and safety profile of the molecule (Li & Easton, 2018).

Hu et al. (2020) conducted a systematic review on 17 biological license applications (BLAs) for biologic–device combinations available with both PFS and AI presentations. This insightful work highlighted that in the majority of applications pharmacokinetic comparability studies were conducted to bridge the two presentations. It was found that most clinical studies were designed to demonstrate bioequivalence. Bioequivalence is assessed based on the area under the concentration–time curve (AUC) and the maximum plasma concentration or peak exposure (C_{max}) for which "the 90% confidence interval for the ratio of the test and reference products should be contained within the acceptance interval of 80.00%–125.00%" (CHMP, 2010).

Depending on the safety, immunogenicity, and tolerability profile of the biological agent, a bioequivalence study is conducted either in healthy volunteers or in patients. Most common trial concepts include a single-dose, parallel group design. Alternatively, especially for products with a high pharmacokinetic variability, a cross-over design may be beneficial to reduce the number of subjects required to demonstrate pharmacokinetic comparability (Zintzaras & Bouka, 1999), as study participants serve as their own control, thus eliminating intersubject variability.

Other successful clinical bridging approaches applied for comparing injection via PFS and AI identified by Hu et al. included analysis of population pharmacokinetic data using sparse pharmacokinetic sampling in phase 3 trials or a comparison of serum trough concentrations (C_{trough}) across different studies (Hu et al., 2020).

Notably, the value of conducting powered bioequivalence studies to bridge from a PFS to an AI presentation is currently debated in the scientific community. In a remarkable commentary to the above review from Hu et al. (2020) on the 17 BLAs for biologic–device combinations with PFS and AI presentations, Lambert (2020) suggests that "formal BE studies should not be required for such bridging studies". The underlying reason for the proposal is that PFS and AI typically contain the same formulation and that the AI "results in a more consistent injection than can be achieved with a PFS." Therefore human factors studies complemented with the established technical device development package (i.e., "device design controls and chemistry, manufacturing, and controls information") and intuitive instructions for use should be sufficient to support effective and safe subcutaneous administration. Lambert explained that his view is further supported by the approval of the AI presentations for products that did not demonstrate bioequivalence to the PFS.

The authors agree with Lambert's assessment. Especially for mAbs available with both an intravenous and a subcutaneous dosing alternative demonstrating formal bioequivalence between two subcutaneous device platforms should not be warranted. These mAbs had previously been shown to exhibit noninferior efficacy with the different administration routes. This is despite a markedly lower C_{max} achieved following subcutaneous as compared to intravenous dosing (Burmester et al., 2014; Davies et al., 2017; Ismael et al., 2012; Mateos et al., 2020). In such cases, adequately designed usability studies along with a comprehensive technical development package should be sufficient to demonstrate safe and effective use of an automated injection device.

Conceptual clinical bridging approach for device technology platforms

When developing subcutaneous dosing alternatives for mAbs, injection devices, such as AIs or OBDSs, are frequently not yet available for use in the pivotal trial and thus subsequently have to be introduced as a lifecycle management following initial launch of the molecule. As described, the underlying bridging program typically includes the demonstration of pharmacokinetic comparability between the PFS or vial presentation injected with a HHS or IS in the pivotal trial with the automated AI or OBDS, respectively. Also, when utilizing the same automated device platform across different projects, a dedicated pharmacokinetic trial is conducted for each mAb (see Fig. 3.1).

This mAb-specific device bridging approach is taken to identify any clinically relevant molecule-specific differences in the absorption from the subcutaneous tissue into the systemic circulation that may depend on the injection procedure with either device technology. The conduct of pharmacokinetic comparability studies for each mAb utilizing the same automated device platform can delay availability of the device and be cost and resources intensive. This is especially true if conducting the pharmacokinetic comparability study in healthy volunteers is precluded due to the safety profile of an individual mAb (Ahmed et al., 2020). Recruiting patients willing to be enrolled into single-dose or cross-over pharmacokinetic studies poses a challenge from an operational perspective, which can result in a prolonged enrollment period (Karakunnel et al., 2018), thus causing delay in device approval.

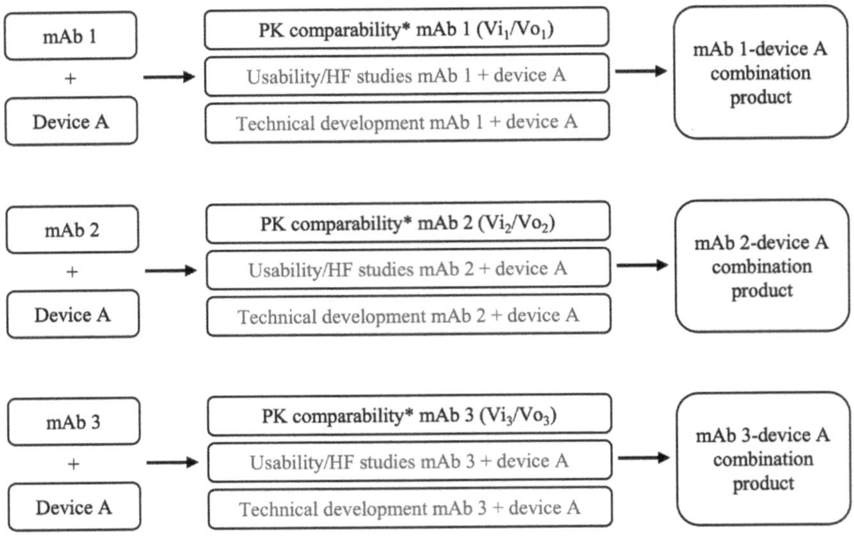

Figure 3.1 Illustration summarizing the current monoclonal antibody–device combination product qualification requirements. *PK (pharmacokinetic) comparability of manual administration applied in the pivotal study (i.e., prefilled syringe or handheld syringe/infusion set) and automated administration using the device (i.e., autoinjectors or on-body delivery systems); assumption: monoclonal antibody completed pivotal study in respective indication(s) using manual administration. *HF*, human factors; *mAb*, monoclonal antibody; *Vi*, viscosity; *Vo*, volume.

In the following, a concept on how to leverage insights generated on the pharmacokinetic comparability between the administration procedure in the pivotal trial and an automated subcutaneous device platform is introduced. The aim is to accelerate the development of mAb–device combinations by omitting the need to repeat pharmacokinetic comparability studies by cross-referencing clinical data from other molecules using the same device platform. Accordingly, clinical device development would be conducted within its operational volume and viscosity range validated by using a "model mAb" (see Fig. 3.2). This approach would replace the need for generating molecule-specific pharmacokinetic data for eligible new mAb formulations applying the same injection device platform within its qualified ranges (see Fig. 3.1).

Molecule-independent clinical device bridging concept

The underlying assumption for the conceptual molecule-independent clinical device bridging approach is that the pivotal clinical study for the mAb was not yet conducted with the automated subcutaneous device platform.

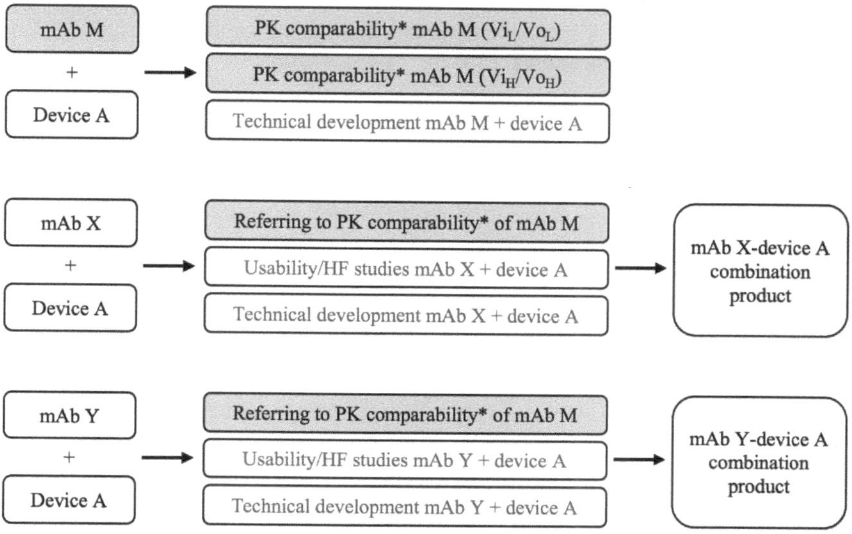

Figure 3.2 Illustration summarizing an alternative conceptual monoclonal antibody–device combination product qualification approach using a "model monoclonal antibody" to demonstrate accuracy of the automated device (i.e., autoinjectors or on-body delivery systems) to deliver the required dose and dosing volume compared to the administration procedure used in the pivotal study (i.e., prefilled syringe or handheld syringe/infusion set). *PK comparability of manual administration applied in the pivotal study (i.e., prefilled syringe or handheld syringe/infusion set) and automated administration using the device (i.e., autoinjectors or on-body delivery systems); assumption: monoclonal antibody completed pivotal study in respective indication(s) using manual administration. *HF*, human factors; *mAb*, monoclonal antibody; *mAb M*, model mAb; *Vi*, viscosity; Vi_H, high viscosity; Vi_L, low viscosity; *Vo*, volume; Vo_H, high volume; Vo_L, low volume.

The platform approach would use a "model mAb" to demonstrate pharmacokinetic comparability for subcutaneous injection via the automated device platform (AI or OBDS) versus the administration procedure used in the pivotal study (PFS or HHS/IS), respectively. A "model mAb" is considered representative to other mAbs with regard to physicochemical and pharmacokinetic properties. Table 3.1 compares the molecule-specific and the conceptual molecule-independent device qualification approach.

Such methodology should allow leveraging related data for other molecules with similar physicochemical and pharmacokinetic properties. Referring to the data generated for the same device platform in other disease areas for a different mAb is expected feasible without compromising patient safety or the scientific integrity of the overall data package.

Table 3.1 Comparison of the molecule-specific and the conceptual molecule-independent device qualification approach.

Activity/data	Molecule-specific device qualification approach	Molecule-independent device qualification approach
Pharmacokinetic comparability study	Performed with each individual mAb	Performed using "model mAb" covering operational viscosity and volume ranges of device platform
	Usability/HF studies	Follows the standard development approach and is performed on each individual study drug intended to be paired with the device combination product/ drug–device combination product
Technical development of mAb–device combination product		

HF, human factors; *mAb*, monoclonal antibody.

Subcutaneous device platforms in scope

AI and OBDS platforms are sterile, nonpyrogenic, single-use, disposable devices that are indicated for subcutaneous injection of prescribed liquid medicines by the target users. Devices may include a visual gage or window indicating injection progress during drug delivery. Such platforms are designed to accommodate different fill volumes while maintaining the same principle of operation.

In this chapter, the term AI is used for different low volume injection devices independent of the applied administration procedure. There are two major AI subtypes with respect to how the user initiates the injection process. With the first device type, the needle and medication flow is actuated by the patient or caregiver by applying pressure on the injection site. An example for this platform is the SensoReady device, approved as a biologic–device combination for secukinumab for the treatment of psoriasis, ankylosing spondylitis, and psoriatic arthritis (Cosyntex), a biosimilar of the antiinflammatory agent etanercept (Erelzi), or for ofatumumab for the treatment of multiple sclerosis (Kesimpta). Alternatively, the user presses a push button that actuates the needle into the targeted injection site. The SureClick device is a prototype for this device technology and is, for example, used for evolocumab for the treatment of hyperlipidemia (Repatha), or tocilizumab and etanercept used to treat different autoimmune diseases (Actemra, Enbrel). Refer to respective SmPc for device details (Electronic Medicines Compendium, 2021).

The first marketed large volume OBDS is the SmartDose device that is filled with a 3.5-mL cartridge containing the dosing solution. This device is approved for the delivery of evolocumab as an alternative to the SureClick AI, requiring only one instead of three consecutive injections for a single monthly dose application (Repatha; Electronic Medicines Compendium, 2021). To initiate the injection process, the patient or caregiver attaches the device to the skin via an adhesive backing for attachment. Medication flow is started by pressing a push button that actuates the needle into the targeted injection site.

Table 3.2 illustrates examples for subcutaneous injection equipment frequently used in pivotal trials as well as the device platforms in scope for this chapter and describes major differences between an AI and an OBDS.

Table 3.2 Major differences in injection procedure and equipment between AI and OBDS platforms.

Type	AI platform	OBDS platform
Fill volume	Up to approximately 2 mL[a]	Expected ranges between, for example, 3–10, 10–25, or 25–50 mL[a,b]
Actuation	Medication flow is actuated by the patient or caregiver by either applying pressure on the injection site or by pressing a push button that actuates the needle into the targeted injection site	Patient or caregiver attaches the device to the skin via an adhesive backing for attachment. Medication flow is actuated by the patient or caregiver by pressing a push button that inserts the needle into the targeted injection site
Needle	Integrated needle with automated needle shield	Integrated needle with automatic retraction mechanism for needle safety lockout

AI, autoinjector; *OBDS*, on-body delivery system.
[a]Approximate only.
[b]Higher volume injection devices are currently in development.

Bridging from a lower volume PFS to an AI device is assumed to be more straightforward as compared to bridging from higher volume manual injection by means of a HHS- or IS-assisted injection to an automated OBDS. This is due to the AI usually hosting the PFS within a shell that automates and simplifies the injection. In addition, this very same PFS was mostly already used in the pivotal studies. Differences in the administration procedure will originate from user-dependent preferences and capabilities related to injection duration and to the individual injection angle that is typically around 45 degrees for the PFS (Marx, 2020) and 90 degrees for the AI (De La Peña et al., 2015). Subcutaneous dosing by means of a PFS will yield a comparatively variable injection time, as the user first needs to overcome the activation force to initiate the injection and subsequently applies a flow rate accounting for individual comfort and capabilities.

To help align the injection procedure between PFS and AI in the pivotal study, the patient or caregiver could be instructed to apply the PFS with the same injection angle and rate as foreseen with the automated AI device to the extent feasible. Such an approach, however, requires careful consideration, as allowing for personal freedom in conducting the subcutaneous injection will generate data that reflect the future clinical situation for the use of the PFS. This is especially important as there is evidence that a number of patients and caregivers would still prefer a PFS over a more automated AI device (Vermeire et al., 2018).

The injection conditions for subcutaneous dosing using a HHS or IS for larger dosing volumes are expected to be more variable as compared to utilizing an automated OBDS. First, due to the greater dosing complexity, drug administration in the pivotal trial will have to be conducted by a professional caregiver and established routines will vary from person to person and between different institutions and countries. If not harmonized via a standard operation procedure, some centers will prefer manual injections, while others will conduct the administration via an IS, so that parameters, such as the needle size and injection rate, will differ. The advantage of such a flexible dosing approach is that the data generated reflect the variability in drug administration without an automated device in the future clinical situation. Standardizing the injection procedure and equipment across all countries and study sites requires the manufacturer to provide the materials with detailed dosing instructions globally. This, however, could be a valuable option to facilitate future bridging to an OBDS.

In both cases, for an AI and an OBDS, use of these automated device types will standardize the injection parameters and administer the formulation at a defined rate, injection angle, and depth. Under the assumption that dosing instructions and equipment are harmonized across study sites in the pivotal clinical trial, it is hypothesized that administration conditions with the automated device are comparable to injection using a PFS or HHS/IS. Most likely, the ranges of injection rates, injection depths, and the injection angles in these studies will be broader than those with the automated device. The subcutaneous injection parameters that need to be harmonized between the subcutaneous administration procedure in the pivotal study and the automated injection device to enable the molecule-independent device bridging approach are listed in Table 3.3.

Injection depth

In a comprehensive review on the subcutaneous delivery of mAbs, Viola et al. (2018) identified what the authors refer to as internal and external factors influencing subcutaneous absorption of mAbs. Internal factors include intrinsic subject characteristics, such as body weight, gender, age, or physical activity, as well as physiology characteristics of the subcutaneous tissue, such as catabolic capacity, physicochemical tissue characteristics, and subcutaneous blood and lymph flow. External factors include the drug and formulation characteristics and the mode of administration. Aspects of the mode of administration encompass the injection site, the application of massage and heat, the anesthesia status, and the volume and depth of injection.

Table 3.3 Formulation and drug administration requirements for products referring to data from molecule-independent device bridging study.

Parameter	Formulation and drug administration requirements
Injection site(s) and subcutaneous tissue layer	Same as in proposed MIDBS
Injection time	Within injection time range of proposed MIDBS
Injection volume	Within range of volumes tested in proposed MIDBS
Viscosity of the drug product	Within range of viscosities tested in proposed MIDBS

MIDBS, molecule-independent device bridging study.

Similarly, in their review of 17 BLAs for biologic–device combinations, Hu et al. (2020) identified the injection depth with the AI or PFS and the injection site as potential influencing factors to the outcome of pharmacokinetic comparability studies. This difference might be explained by a different pressure gradient between lymphatics and interstitium that plays an important role in lymphatic uptake (Richter et al., 2012). Taken all these factors together and comparing with the fixed and variable factors when bridging from a PFS or HHS/IS to an automated AI or OBDS, the parameter that is expected to introduce most variability at a given injection site and therefore should to be standardized in the pivotal study is the injection depth.

Eligible biologics and formulations in-scope

It is hypothesized that molecules suitable for cross-referring to the data from the molecule-independent device bridging approach for a selected automated device platform possess a comparable physicochemical and pharmacokinetic profile. Moreover, the AI or OBDS has to contain the same formulation (i.e., including the same excipients at the same quantities) and dosing volume as in the pivotal trials.

Specifically, possible variations in the subcutaneous injection procedure with different device types may impact the absorption rate of a molecule into the bloodstream, which might be reflected by differences in bioavailability, C_{max}, and corresponding t_{max}. Distribution and elimination of the biologic, however, are not expected to be affected by changes in drug administration modalities.

mAbs in scope for the molecule-independent device bridging approach for mAb–device combinations would therefore be characterized by a slow absorption from the subcutaneous tissue into the systemic circulation. The underlying rationale is that in such situations, the release from the subcutaneous space via lymph flow rather than the particularities of the injection method is expected to be the rate limiting factor for C_{max}. As summarized by Ryman and Meibohm (2017), the absorption of mAbs from the subcutaneous tissue into the systemic circulation is primarily via the lymphatic vessels and typically slow, with t_{max} values ranging between approximately 6 and 8 days. With the aim to describe the absorption process of mAbs following subcutaneous injection, Zhao et al. developed a hybrid model formed by coupling the physiologically based absorption process with a conventional pharmacokinetic–pharmacodynamic or pharmacokinetic model associated with intravenous infusion. In their simulations, lymphatic flow rate was the only influential factor to t_{max}. Contribution of clearance of the antibody from the endosomal layer at the injection site was found to be marginal. Furthermore, the authors summarized that different injection volumes might impact lymphatic flow and that changes in the physiological environment at the injection site might be a result of different formulation ingredients. Consequently, the molecule-independent device bridging approach can only be applied when the automated AI or OBDS contains the same formulation and dosing volume applied in the pivotal clinical trial for the respective mAb.

The hypothesis that the particularities of the injection procedure have a limited impact on bioavailability, C_{max}, and t_{max} is supported by an investigation of the pharmacokinetics, safety, and tolerability of secukinumab following varying subcutaneous administration conditions (Bruin et al., 2020). Namely, secukinumab was administered using a variety of delivery systems in healthy volunteers and in psoriasis patients. In the healthy volunteer study, six different administration conditions were assessed to inject 2 mL of a 150 mg/mL solution of secukinumab. The authors compared AI, PFS, and pump systems with manual injection. In addition to the device type, injections differed in needle size, injection depth, injection rate, and the number of injections for each individual dose. It was concluded that use of the six dosing modalities resulted in "very comparable serum concentration-time profiles," including C_{max} and that the median t_{max} for each dosing condition ranged between 4.5 and 7 days. Results from the phase 3 study in psoriasis patients, in which multiple doses of secukinumab were administered using a PFS either with two 1-mL injections or with one 2-mL injection confirmed the findings from the healthy volunteers study and the mean serum concentration—time profiles of secukinumab were comparable independent on whether the dose was administered with one versus two injections.

Considering that an automated AI or OBDS device platform will deliver the same formulation with the same dose and dosing volume as the PFS or vial presentation in the pivotal trial, the impact of the device technology on the pharmacokinetic profile of a mAb that exhibits a slow absorption rate is expected to be marginal.

Injection sites

It has been observed for a number of biological molecules that the rate and extent of absorption from the subcutaneous tissue into the systemic circulation can depend on the injection site (Beshyah et al., 1991; Bruin et al., 2020; Kagan et al., 2012; Möbus et al., 2018). The direction of this difference depends on the individual mAb. Bruin et al. (2020) noted a slightly elevated exposure combined with an earlier t_{max} following subcutaneous injection of secukinumab into the thigh compared to the abdomen with geometric mean ratio for thigh to abdomen of 1.19 (AUC) and a 90% confidence interval (CI) of 0.97−1.46. In a study with bococizumab, subcutaneous injection into the abdomen, thigh, or upper arm resulted in comparable pharmacokinetic profiles (Wang et al., 2017). The observed injection site-dependent variability in the subcutaneous bioavailability may be rationalized by differences in blood and lymph flow or in a different degree of presystemic or systemic catabolism in draining lymphatics and blood vessels (Richter & Jacobsen, 2014) following thigh, arm, or abdominal injection.

An approach to assess the relative impact of the device type versus the injection site on the pharmacokinetics of a biological molecule is to conduct a device bridging trial that compares manual versus automated subcutaneous injection at different injection sites. Martin et al. (2021) compared the pharmacokinetics of benralizumab following injection either by means of an automated AI or a PFS. Injection sites were the abdomen, the thigh and the upper arm. Overall, and independent of the site of administration, exposure was similar between the AI and the PFS. Geometric mean ratios for AUC_{last} and AUC_{inf} (AI vs PFS) (90% CI) were 0.93 (0.87−0.99) and 0.95 (0.88−1.02), respectively. It was found that exposure following injection into the thigh was approximately 15%−30% greater than for abdomen or upper arm administration. Namely, for comparisons of the thigh to other injection sites, the upper limit of the 90% CIs of the geometric mean ratios was greater than the reference range of 125%, and the lower limit of 90% CIs was greater than 100%.

Under the assumption that the impact of the injection site on the pharmacokinetic profile of a mAb is independent of the administration device, an assessment of different injection sites would not be part of the conceptual molecule-independent device bridging approach. Here, only injection into the abdominal tissue would be studied. Use of the automated device for additional body areas, such as the thigh or upper arm, would only be feasible for mAbs using the device platform in case available data with the previous administration procedure revealed no clinically relevant differences in the pharmacokinetics for either injection site.

Selection of "model mAb"

To account for the challenges in recruiting patients into a pharmacokinetic study and to limit the number of participants, the molecule-independent device bridging study should ideally be conducted in healthy volunteers, a population with expected lower pharmacokinetic variability. To mimic a typical mAb, the "model mAb" for this trial should be characterized by a pharmacokinetic profile in the range of other mAbs in scope for cross-referring to the data generated. As a change in injection conditions may impact the release from the interstitial tissue, the absorption rate should be comparable to that of other mAbs foreseen to be delivered via the novel device technology.

The use of a "model mAb" is considered justified, as the intent of the bridging study is to confirm a comparable pharmacokinetic profile between two administration procedures (e.g., manual injection using PFS or HHS/IS vs automated AI or OBDS), injecting the same formulation at the same dose and dosing volume. The study should utilize a mAb that has been demonstrated safe and well tolerated up to a high dosing volume, displays dose-proportional pharmacokinetics over the desired injection volume range, and can be administered to healthy volunteers.

Conceptual molecule-independent device bridging study

There are a number of assumptions underlying the applicability of the molecule-independent device bridging study. The mAb used in the pivotal study is the same as in the mAb–device combination containing the same active molecule and the same formulation and excipient composition. The difference is the presentation in a PFS or vial versus an automated AI or OBDS. To the extent possible, the performance parameters of the biologic–device combination have to mimic the parameters of the administration procedure in the pivotal study performed with a PFS or HHS/IS. Specific technical device development work, such as selection of materials, compatibility, and stability with the specific study drug, as well as technical specifications and functionality tests, would follow current Health Authorities' guidance and are out of scope for this chapter.

Design of the conceptual molecule-independent device bridging study

To mimic drug administration in pivotal studies and support bridging to suitable mAbs using the same device platform, the following parameters need to be fixed in the bridging trial (Table 3.3). In both study arms, the drug is administered into the same injection site(s) and target tissue layer. To the extent possible, a comparable needle length and

Table 3.4 Treatments in molecule-independent device bridging study.

Treatment	Injection procedure		Viscosity	Volume
	Volumes up to 2 mL*	**Volumes greater than 2 mL***		
1: Test_Vi_HVo_H	AI	OBDS	High	High
2: Test_Vi_LVo_L	AI	OBDS	Low	Low
3: Control_Vi_HVo_H	PFS	HHS/IS	High	High
4: Control_Vi_LVo_L	PFS	HHS/IS	Low	Low

Note: The device could either be an AI or an OBDS.
AI, autoinjectors; *IS*, infusion set; *OBDS*, on-body delivery systems; *PFS*, prefilled syringe.
*Approximate only.

gage between PFS or HHS/IS and the automated AI or OBDS platform is to be used. The injection time with the automated device needs to be in the range of that in the pivotal study. The injection volumes and viscosities tested in the device bridging study need to bracket the expected range of volumes and viscosities of formulations foreseen to use the device platform.

The conceptual bridging trial could be designed as an open-label, parallel-group study. Participants would be randomized to receive a single subcutaneous injection into the abdomen either via the automated AI or OBDS platform (test arm) or via the administration procedure in the pivotal clinical trial using a PFS or HHS/IS (control arm). This parallel-group approach, as opposed to a cross-over design, would account for a comparably long half-life of up to approximately 30 days for a mAb (Ovacik & Lin, 2018).

The selected "model mAb" would be formulated with two viscosities and two injection volumes to validate the injection device functionality ranges by testing the two extremes (low viscosity/low volume Vi_LVo_L and high viscosity/high volume Vi_HVo_H) (Table 3.4).

Summary and outlook

The pressure on healthcare costs and resources is increasing and at the same time the demand for customer-centric product presentations that allow a flexible care setting for suitable therapeutics, where patients and healthcare providers can select the place of drug administration, has markedly increased. This change has been accelerated in the COVID-19 pandemic and will continue to force manufacturers to develop safe and effective biotherapeutics that are also easy to administer ideally in an environment preferred by the individual patient. Subcutaneous dosing alternatives have already largely facilitated parenteral dosing of biotherapeutics, and ready-to-use injection devices that can be handled by patients and their lay caregivers further facilitate parenteral drug administration in the patient's home.

Bridging to novel device technologies that have not yet been an integral part of the pivotal clinical trials for a novel biological medicine has been established for AI devices since many years. Manufacturers are mandated to demonstrate safe and effective handling in human factor studies including simulated use assessments, and, depending on the device type and patient population, in some cases real use studies. Moreover, to guarantee that the changed injection procedure does not result in clinically relevant changes in systemic exposure of the molecule, pharmacokinetic comparability studies are typically part of the data package supporting registration. Independent of whether a subcutaneous AI platform has been registered, these studies are usually repeated for each individual molecule.

Today, larger volume OBDSs are entering the market, supporting home- and self-administration of high-dose biopharmaceutics. Due to the relative novelty of the underlying technologies, timely availability for pivotal trials of corresponding biotherapeutics is still unlikely, at least initially. The availability of a technology platform that could be applied to different molecules and is upgraded over time could possibly enable launching the device together with a novel biotherapeutic in the future. In addition, use of device platforms can offer a number of benefits including economy of scale, increased familiarity with the device when used with a number of biotherapeutics, and the opportunity to cross-reference clinical data from one molecule to another.

To accelerate the development of automated device technologies for individual mAbs, the described molecule-independent device bridging approach using a "model mAb" to demonstrate pharmacokinetic comparability would be an opportunity provided its acceptability by regulators. The prerequisite for an eligible mAb to refer to the pharmacokinetic data comparing the administration methodology in the pivotal trial to that with the automated device will be that the same formulation, the same dose and dosing volume is delivered both ways. The evolving database, including delivery of large volume biotherapeutics, is expected to further inform the concept from this point forward.

References

Ahmed, M. A., Patel, C., Drezner, N., Helms, W., Tan, W., & Stypinski, D. (2020). Pivotal considerations for optimal deployment of healthy volunteers in oncology drug development. *Clinical and Translational Science*, *13*(1), 31–40. Available from https://doi.org/10.1111/cts.12703.

Beshyah, S. A., Anyaoku, V., Niththyananthan, R., Sharp, P., & Johnston, D. G. (1991). The effect of subcutaneous injection site on absorption of human growth hormone: Abdomen versus thigh. *Clinical Endocrinology*, *35*(5), 409–412. Available from https://doi.org/10.1111/j.1365-2265.1991.tb03557.x.

Bittner, B., & Schmidt, J. (2021). Subcutaneous drug delivery devices—Enablers of a flexible care setting. In E. Chappel (Ed.), *Drug delivery devices and therapeutic systems* (pp. 159–179). Elsevier. Available from https://doi.org/10.1016/b978-0-12-819838-4.00021-3.

Brand-Schieber, E., Munjal, S., Kumar, R., Andre, A. D., Valladao, W., & Ramirez, M. (2016). Human factors validation study of 3 mg sumatriptan autoinjector, for migraine patients. *Medical Devices: Evidence and Research*, *9*, 131–137. Available from https://doi.org/10.2147/MDER.S105899.

Bruin, G., Hockey, H. U. P., La Stella, P., Sigurgeirsson, B., Fu, R., Patekar, M., Charef, P., Woessner, R., & Boutouyrie-Dumont, B. (2020). Comparison of pharmacokinetics, safety and tolerability of secukinumab administered subcutaneously using different delivery systems in healthy volunteers and in psoriasis patients. *British Journal of Clinical Pharmacology*, *86*(2), 338–351. Available from https://doi.org/10.1111/bcp.14155.

Burmester, G. R., Rubbert-Roth, A., Cantagrel, A., Hall, S., Leszczynski, P., Feldman, D., Rangaraj, M. J., Roane, G., Ludivico, C., Lu, P., Rowell, L., Bao, M., & Mysler, E. F. (2014). A randomised, double-blind, parallel-group study of the safety and efficacy of subcutaneous tocilizumab versus intravenous tocilizumab in combination with traditional disease-modifying antirheumatic drugs in patients with moderate to severe rheumatoid arthritis (SUMMACTA study). *Annals of the Rheumatic Diseases*, *73*(1), 69–74. Available from https://doi.org/10.1136/annrheumdis-2013-203523.

CHMP. (2010). Guideline on the investigation of bioequivalence, Committee for Medicinal Products for Human use (CHMP). https://www.ema.europa.eu/en/documents/scientific-guideline/guideline-investigation-bioequivalence-rev1_en.pdf.

CHMP. (2019). Guideline on the quality requirements for drug-device combinations, Committee for Medicinal Products for Human Use (CHMP). https://www.ema.europa.eu/en/documents/scientific-guideline/draft-guideline-quality-requirements-drug-device-combinations_en.pdf.

Clark, S., & Forkey, P. E. D. (2018). Clinical actual use human factors testing. *Proceedings of the International Symposium on Human Factors and Ergonomics in Health Care*, 7(1), 152–154. Available from https://doi.org/10.1177/2327857918071039.

Davies, A., Merli, F., Mihaljević, B., Mercadal, S., Siritanaratkul, N., Solal-Céligny, P., Boehnke, A., Berge, C., Genevray, M., Zharkov, A., Dixon, M., Brewster, M., Barrett, M., & MacDonald, D. (2017). Efficacy and safety of subcutaneous rituximab versus intravenous rituximab for first-line treatment of follicular lymphoma (SABRINA): A randomised, open-label, phase 3 trial. *The Lancet Haematology*, *4*(6), e272–e282. Available from https://doi.org/10.1016/S2352-3026(17)30078-9.

De La Peña, A., Yeo, K. P., Linnebjerg, H., Catton, E., Reddy, S., Brown-Augsburger, P., Morrow, L., & Ignaut, D. A. (2015). Subcutaneous injection depth does not affect the pharmacokinetics or glucodynamics of insulin lispro in normal weight or healthy obese subjects. *Journal of Diabetes Science and Technology*, *9*(4), 824–830. Available from https://doi.org/10.1177/1932296815573865.

Domańska, B., Stumpp, O., Poon, S., Oray, S., Mountian, I., & Pichon, C. (2018). Using patient feedback to optimize the design of a certolizumab pegol electromechanical self-injection device: Insights from human factors studies. *Advances in Therapy, 35*, 100–115. Available from https://doi.org/10.1007/s12325-017-0645-1.

Electronic Medicines Compendium. (2021). https://www.medicines.org.uk/emc.

FDA. (2013). Technical considerations for pen, jet, and related injectors intended for use with drugs and biological products, FDA guidance for industry. https://www.fda.gov/media/76403/download.

FDA. (2016). Human factors studies and related clinical study considerations in combination product design and development, draft FDA guidance for industry. https://www.fda.gov/media/96018/download.

FDA. (2019). Bridging for drug-device and biologic-device combination products, draft FDA guidance for industry. https://www.fda.gov/media/133676/download

Fettner, S., Mela, C., Wildenhahn, F., Tavanti, M., Wells, C., Douglass, W., & Mallalieu, N. L. (2019). Evidence of bioequivalence and positive patient user handling of a tocilizumab autoinjector. *Expert Opinion on Drug Delivery, 16*(5), 551–561. Available from https://doi.org/10.1080/17425247.2019.1604678.

Hu, P., Wang, J., Florian, J., Shatzer, K., Stevens, A. M., Gertz, J., Ji, P., Huang, S. M., Zineh, I., & Wang, Y. M. C. (2020). Systematic review of device parameters and design of studies bridging biologic-device combination products using prefilled syringes and autoinjectors. *AAPS Journal, 22*(52). Available from https://doi.org/10.1208/s12248-020-0433-8.

Ismael, G., Hegg, R., Muehlbauer, S., Heinzmann, D., Lum, B., Kim, S. B., Pienkowski, T., Lichinitser, M., Semiglazov, V., Melichar, B., & Jackisch, C. (2012). Subcutaneous versus intravenous administration of (neo)adjuvant trastuzumab in patients with HER2-positive, clinical stage I-III breast cancer (HannaH study): A phase 3, open-label, multicentre, randomised trial. *The Lancet Oncology, 13*(9), 869–878. Available from https://doi.org/10.1016/S1470-2045(12)70329-7.

Kagan, L., Turner, M. R., Balu-Iyer, S. V., & Mager, D. E. (2012). Subcutaneous absorption of monoclonal antibodies: Role of dose, site of injection, and injection volume on rituximab pharmacokinetics in rats. *Pharmaceutical Research, 29*(2), 490–499. Available from https://doi.org/10.1007/s11095-011-0578-3.

Karakunnel, J. J., Bui, N., Palaniappan, L., Schmidt, K. T., Mahaffey, K. W., Morrison, B., Figg, W. D., & Kummar, S. (2018). Reviewing the role of healthy volunteer studies in drug development. *Journal of Translational Medicine, 16*(1). Available from https://doi.org/10.1186/s12967-018-1710-5.

Lambert, W. J. (2020). Why do the majority of submissions for bridging from a prefilled syringe to an autoinjector include bioequivalence studies in order to demonstrate comparability? *AAPS Journal, 22*(3). Available from https://doi.org/10.1208/s12248-020-00453-0.

Lange, J., & Nemeth, T. (2018). Formative usability evaluation of a fixed-dose pen-injector platform device. *Medical Devices: Evidence and Research, 11*, 105–112. Available from https://doi.org/10.2147/MDER.S159733.

Li, Z., & Easton, R. (2018). Practical considerations in clinical strategy to support the development of injectable drug-device combination products for biologics. *MAbs, 10* (1), 18–33. Available from https://doi.org/10.1080/19420862.2017.1392424.

Maropoulos, P. G., & Ceglarek, D. (2010). Design verification and validation in product lifecycle. *CIRP Annals - Manufacturing Technology, 59*(2), 740–759. Available from https://doi.org/10.1016/j.cirp.2010.05.005.

Martin, U. J., Fuhr, R., Forte, P., Barker, P., Axley, M. J., Aurivillius, M., Yan, L., & Roskos, L. (2021). Comparison of autoinjector with accessorized prefilled syringe for benralizumab pharmacokinetic exposure: AMES trial results. *Journal of Asthma*, *58*(1), 93–101. Available from https://doi.org/10.1080/02770903.2019.1663428.

Marx, M. (2020). Evidence-based guidance for self-administration of injectable contraception. *Journal of Midwifery and Women's Health*, *66*(1), 108–112. Available from https://doi.org/10.1111/jmwh.13190.

Mateos, M. V., Nahi, H., Legiec, W., Grosicki, S., Vorobyev, V., Spicka, I., Hungria, V., Korenkova, S., Bahlis, N., Flogegard, M., Bladé, J., Moreau, P., Kaiser, M., Iida, S., Laubach, J., Magen, H., Cavo, M., Hulin, C., White, D., & Usmani, S. Z. (2020). Subcutaneous versus intravenous daratumumab in patients with relapsed or refractory multiple myeloma (COLUMBA): A multicentre, open-label, non-inferiority, randomised, phase 3 trial. *The Lancet Haematology*, *7*(5), e370–e380. Available from https://doi.org/10.1016/S2352-3026(20)30070-3.

MHRA (2021). Guidance on applying human factors and usability engineering to medical devices including drug-device combination products in Great Britain, Medicines & Healthcare products Regulatory Agency (MHRA), Version 2.0. https://assets.publishing.service.gov.uk/government/uploads/system/uploads/attachment_data/file/970563/Human-Factors_Medical-Devices_v2.0.pdf.

Möbus, V., Mahlberg, R., Janni, W., Tomé, O., Marmé, F., Forstbauer, H., Reimer, T., von der Assen, A., Reinisch, M., Lorenz, R., Schmatloch, S., Schmid, M., Sinn, B., Klutinus, N., Stickeler, E., Untch, M., Seiler, S., Burchardi, N., von Minckwitz, G., & Loibl, S. (2018). Pharmacokinetic results of a subcutaneous injection of trastuzumab into the thigh versus into the abdominal wall in patients with HER2-positive primary breast cancer (BC) treated within the neo-/adjuvant GAIN-2 study. American Association for. *Cancer Research*, *78*(4). Available from https://doi.org/10.1158/1538-7445.SABCS17-P5-20-09.

Newswanger, B., Prestrelski, S., & Andre, A. D. (2019). Human factors studies of a prefilled syringe with stable liquid glucagon in a simulated severe hypoglycemia rescue situation. *Expert Opinion on Drug Delivery*, *16*(9), 1015–1025. Available from https://doi.org/10.1080/17425247.2019.1653278.

Ovacik, M., & Lin, K. (2018). Tutorial on monoclonal antibody pharmacokinetics and its considerations in early development. *Clinical and Translational Science*, *11*(6), 540–552. Available from https://doi.org/10.1111/cts.12567.

Richter, W. F., Bhansali, S. G., & Morris, M. E. (2012). Mechanistic determinants of biotherapeutics absorption following SC administration. *AAPS Journal*, *14*(3), 559–570. Available from https://doi.org/10.1208/s12248-012-9367-0.

Richter, W. F., & Jacobsen, B. (2014). Subcutaneous absorption of biotherapeutics: Knowns and unknowns. *Drug Metabolism and Disposition*, *42*(11), 1881–1889. Available from https://doi.org/10.1124/dmd.114.059238.

Rojas, K. M., Cosler, L., & Santos, D. L. (2019). A narrative review of FDA human factors validation requirement: The needs of key stakeholders and proposal of an industry (human factors service providers) maturity assessment tool. *Proceedings of the International Symposium on Human Factors and Ergonomics in Health Care*, *8*(1), 222–233. Available from https://doi.org/10.1177/2327857919081056.

Ryman, J. T., & Meibohm, B. (2017). Pharmacokinetics of monoclonal antibodies. CPT: Pharmacometrics and systems. *Pharmacology*, *6*(9), 576–588. Available from https://doi.org/10.1002/psp4.12224.

Travanty, M. N., Calawa, B., Shalaby, W. S., Jozwiakowski, M. J., & Haraldsen, K. B. (2018). Development and usability of a new subcutaneous auto-injector device to administer hydroxyprogesterone caproate to reduce the risk of recurrent preterm birth. *Medical Devices: Evidence and Research*, *11*, 241–252. Available from https://doi.org/10.2147/MDER.S157114.

Vermeire, S., D'heygere, F., Nakad, A., Franchimont, D., Fontaine, F., Louis, E., Van Hootegem, P., Dewit, O., Lambrecht, G., Strubbe, B., & Baert, F. (2018). Preference for a prefilled syringe or an auto-injection device for delivering golimumab in patients with moderate-to-severe ulcerative colitis: A randomized crossover study. *Patient Preference and Adherence*, *12*, 1193–1202. Available from https://doi.org/10.2147/PPA.S154181.

Viola, M., Sequeira, J., Seiça, R., Veiga, F., Serra, J., Santos, A. C., & Ribeiro, A. J. (2018). Subcutaneous delivery of monoclonal antibodies: How do we get there? *Journal of Controlled Release*, *286*, 301–314. Available from https://doi.org/10.1016/j.jconrel.2018.08.001.

Wang, E. Q., Plotka, A., Salageanu, J., Sattler, C., & Yunis, C. (2017). Pharmacokinetics and pharmacodynamics of bococizumab, a monoclonal antibody to PCSK9, after single subcutaneous injection at three sites [NCT 02043301]. *Cardiovascular Therapeutics*, *35*(5). Available from https://doi.org/10.1111/1755-5922.12278.

Zintzaras, E., & Bouka, P. (1999). Bioequivalence studies: Biometrical concepts of alternative designs and pooled analysis. *European Journal of Drug Metabolism and Pharmacokinetics*, *24* (3), 225–232. Available from https://doi.org/10.1007/BF03190024.

CHAPTER 4

Enabling a flexible care setting in oncology—how to realize home- and self-administration of biotherapeutics in cancer care

Introduction

The underlying rationale of formulation and device lifecycle management for biotherapeutics is to reduce drug administration complexity and, thus, to enable and facilitate dosing in a flexible care setting. In such a setting, patients and their lay caregivers can select the place of drug administration according to individual preferences and capabilities. This includes dosing in the clinic, a physician's office, a community center or the patient's home, and possibly patient self-administration. Especially, due to the increasing cost and resource pressure on the global healthcare system and even more forcefully in light of the COVID-19 pandemic, the value of enabling cancer care outside of a healthcare institution is broadly recognized (Bhatla & Singhal, 2020; Laughlin et al., 2020).

Nonsupervised parenteral drug administration is already realized in many disease areas. In indications such as rheumatoid arthritis (RA), asthma, or multiple sclerosis (MS), a variety of biotherapeutics are self-administered subcutaneously making use of low-volume dosing solutions and prefilled syringes (PFS), prefilled autoinjector (AI) or pen (PFP) devices (Bittner et al., 2018). Currently, the situation is different for biologics, such as monoclonal antibodies (mAbs), used in the treatment of cancer. Here, safety profiles are frequently characterized by infusion-related reactions (Rombouts et al., 2020), occurring predominantly during the initial treatment cycles and requiring close monitoring of the patient (Roselló et al., 2017). With the predominant administration route for biologics in oncology being intravenous infusion, the inherent complex administration procedure further complicates drug administration outside of a healthcare institution. More recently, less invasive subcutaneous dosing alternatives have been approved for mAbs in oncology (Bittner et al., 2018).

Formulation and Device Lifecycle Management of Biotherapeutics
DOI: https://doi.org/10.1016/B978-0-12-823741-0.00006-3
© 2022 Elsevier Inc.
All rights reserved.

Such drug delivery advances create an opportunity to facilitate a flexible care setting for biologics with an acceptable safety and tolerability profile and for patients and their lay caregivers who are willing and adept of parenteral drug administration without the support of a healthcare provider.

The aim of this chapter is to assess the barriers to realizing a flexible care setting for biotherapeutics in cancer care. Focus is on how to address the expected needs and opportunities for patients and healthcare providers as well as on current drug administration—related barriers. Case studies of biologics with a self-administration label outside of oncology serve as precedents for proposing the next steps required to facilitate home- and self-administration for eligible medications and cancer patients.

The importance of simplifying drug delivery of biotherapeutics in oncology

Today, treatment and management of cancer poses significant financial implications to patients and the overall healthcare system (Hofmarcher et al., 2020). Key drivers for the high financial burden on cancer care are the costs of the biological medication itself (Hernandez et al., 2018), country-dependent healthcare provider fees related to complex dosing regimens (Dowd & Laugesen, 2020), and resource requirements associated with hospital and outpatient facilities, especially for parenteral drug administration (Basic et al., 2020; Tetteh & Morris, 2014). The growing use of novel parenteral biologics is expected to further add to the pressure on the healthcare system.

At the same time, with the advances in oncology medications over the recent years, many patients are now living with cancer like a chronic disease (Phillips & Currow, 2010). Thus, supporting patients in becoming more independent with simplified drug delivery solutions is increasingly gaining importance. Consequently, in-line with other chronic conditions, formulation and device-based enablers for treatment management in the home setting need to be provided.

The need for biotherapeutics to be administered in a flexible care setting has drastically gained urgency with the onset of the COVID-19 pandemic and the first lockdowns back in 2020 to manage restricted access to clinics or physician's offices and to offer continuous uninterrupted treatment. In a situation with global travel restrictions and the need to protect vulnerable patients from getting infected with the virus (Al-Shamsi et al., 2020), delivery of care in a healthcare institution can become a challenge for patients and their caregivers (Hanna et al., 2021). Options to administer a medication at home or close to the patient's home are turning into a must. Consequently, the development of supporting drug delivery modalities and of measures that ensure safe and accurate parenteral administration of biotherapeutics without the direct supervision of a healthcare provider is mandated. Product optimizations that enable moving cancer treatment from a healthcare institution-based to a flexible care setting are a key activity manufacturers now embark in already early during the lifecycle of a biologic drug, even in oncology.

It is important to note that still not all cancer patients will be eligible for drug administration outside of a healthcare institution. Also, home-administration needs to be supported by an appropriate safety and tolerability profile established in clinical trials and possibly dependent on the tolerability of an individual patient.

How to unfold the full value of a flexible care setting in cancer care?

To ascertain parenteral administration of biological treatments in a flexible care setting, a number of challenges will need to be appropriately addressed. Key focus areas in this context comprise inherent patient and healthcare provider barriers, as well as drug delivery-related treatment barriers.

Patient barriers and how to overcome them

In theory, from a cancer patient perspective, the option to receive their parenteral treatment at home or in a community center close to their home saves significant traveling and waiting time in the clinic and thus reduces the risk of infections. Patients would overall gain more independence and flexibility in arranging their daily life, an aspect that could be especially important for patients and their lay caregivers who need to take time off from work for receiving the treatment. These factors will likely translate into practicable home dosing without negatively affecting quality-of-life compared to standard ambulatory hospital care (Cool et al., 2019).

While it is anticipated that many patients will appreciate the option of home-administration (Tralongo et al., 2011), individual patients may sense uncertainty with dosing in an uncontrolled setting, especially due to the dosing complexity inherent to parenteral administration and possible side effects. In addition, patients may feel isolated as a result of at-home care (Biagioli et al., 2020), missing the direct relationship with their doctor and the social contact with other patients as experienced in infusion rooms (Hwang et al., 2020).

To overcome the challenges innate to parenteral home-administration for patients, a strong communication network between patient, caregiver, and the healthcare provider is mandated (Tralongo et al., 2011). Only patients who tolerated initial treatment cycles and demonstrated capability of dosing without the supervision of a healthcare provider would be eligible for continued treatment at home. Treatment initiation by an experienced physician like in other disease areas (Chaplin, 2020) is expected to help alleviate anxiety and understand the dosing procedure. The patient or a member of his or her family needs to be properly trained in how to administer the medication (Alshamrani et al., 2020), for example, administering the correct dosage and adhering to the dosing procedure. In addition, easily comprehensible education about the importance of complying with the dosing regimen, as well as about the consequences of poor adherence has to be provided prior to continuing dosing at home. Patients and caregivers require intelligible guidance regarding treatment-related adverse events (Bordonaro et al., 2014) to understand how to handle potential side effects. Adherence aids, such as telemedicine tools to remotely monitor patients, can be supportive (Uzzo et al., 2020), and should maintain a link back to the oncologist, so patients are reassured that the treatment is still being supervised by an expert.

Healthcare provider barriers and how to overcome them

It is supposed that from a healthcare provider perspective, shifting parenteral drug administration for eligible patients to a flexible care setting will reduce the infusion-related burden for their institutions (Dent et al., 2019). Dosing in the home setting will increase resource availability for individual patients, thus improve quality of services and ultimately the reputation of their center. Assuming adequate procedures to ensure adherence to treatment outside of a controlled environment is put in place, home-administration would also reduce the risk for infections for the vulnerable cancer patient population in times of a global pandemic, while at the same time ensuring continued application of a safe and efficacious medication (Uzzo et al., 2020).

Inherent to the role of the healthcare provider to warrant that their patient receives the best possible treatment and feels comfortable with the dosing procedures, the perceived barriers for patients also need to be overcome for healthcare providers to increase comfort in prescribing biological medication for home-administration. An additional key barrier to a flexible care setting from an oncologist's perspective could be the loss of direct patient interactions (Barrios et al., 2020), while still being liable for patient treatment.

To maintain quality of cancer care in a noncontrolled setting, an initial upfront setup investment will be required, including adequate education and training of nurses and administrative employees involved in the conduct and management of home healthcare services (Vaartio-Rajalin et al., 2020). Such mobile services include home infusion, nursing visits, and advanced practice provider or physician visits (Handley & Bekelman, 2019). Related expenditures supposedly will largely pay back with the increasing need for decentralized cancer care during and beyond COVID-19 pandemic, as at-home dosing is no longer a nice to have, but a necessity.

To further increase the comfort of oncologists in parenteral home-administration of biologicals, manufacturers and regulators need to establish concepts and pathways to support physicians in identifying the most appropriate patient population. Especially for biotherapeutics that exhibit infusion-related reactions during initial treatment cycles, dosing should be initiated and supervised by a healthcare provider in the clinic or physician's office, as it is common practice in other disease areas. This way, oncologists can engage into a training exercise together with the patients. During this process, the therapy is first administered by a healthcare professional, then the patient or caregiver administers him- or herself under supervision. This is followed by close monitoring during a nonobserved period until the patient, lay caregiver and treating physician feel comfortable with self-administering alone. Moreover, following this controlled treatment initiation phase, physicians will have observed the tolerability profile of a given biological treatment at an individual patient basis.

Current expert discussions during the COVID-19 pandemic focus on expanding access to subcutaneous treatments via home-administration (Alshamrani et al., 2020). Suggestions on how to get there include less frequent dosing regimens (Malard & Mohty, 2020) that reduce the number of clinic visits over the treatment period, telemedicine to ensure and monitor compliance in a remote setting, or in-home services (Foley et al., 2020) to support patients in procedures related to dosing and general care.

Electronic collection of patient-generated health data can improve transparency of at-home administration through remote monitoring, but the lack of supporting reimbursement mechanisms still remains a barrier for advancement of connected health (Lavallee et al., 2020). Ideally, healthcare providers would receive compensation for the time required for patient education to self-administer, analog to the subcutaneous or intravenous administration of a treatment by the oncologist or staff. To secure reimbursement of healthcare providers for additional work inherent to remote monitoring of patients, a value-based evaluation of home-administration is required to demonstrate the benefit of decentralization of care and the resources saved. Along these lines, the Centers for Medicare and Medicaid Services, the agency within the United States Department of Health and Human Services that administers the nation's major healthcare programs, proposed a rule to expand value-based payment to in-home healthcare that would make at-home infusion available to Medicare beneficiaries. This approach allows patients to receive a higher level of care in their home environment without having to go to the hospital or a provider's office (Waddill, 2019).

In fact, and in light of the COVID-19 pandemic, the fee-for-service payment system in which providers are reimbursed for the complexity of dosing, including related tests, procedures, and in-clinic patient visits, is further challenged in times of social distancing with protocols limiting the number of patients in office settings. It is therefore expected that instead of reimbursing healthcare providers for the individual services provided, payment for a population of patients with, for example, global budget payment models, would avoid the COVID-19-dependent revenue shortages (Meuse, 2020).

Treatment barriers and how to overcome them

In the following, treatment barriers are defined as logistical barriers regarding how the treatment is prepared, stored, and administered to the patient.

Especially when it comes to complex and invasive parenteral administration of biotherapeutics (Bittner et al., 2012), the overall product presentation and the dosing procedure play a pivotal role as enablers for a flexible care setting. Logistically, patients' homes may not be set-up to store medication, and adequate refrigeration to prevent cool chain breakage from the pharmacy to home may not be feasible without investment into equipment enabling tight temperature control. Disposal of needles and potent residual medication can be also challenging (Makki et al., 2019), especially for patients living in rural areas and not in close proximity to a hospital or pharmacy. Parenteral administration can be complicated with regard to dose calculation, numerous preparatory steps, and invasive infusion procedures. The need for complex drug administration equipment bears a notable risk for inadequate dosing. In addition, especially when using partial vials for weight-adjusted dosing regimens, the generation of costly drug wastage needs to be taken into consideration (Lindsey et al., 2018). Combination therapy, in which two or more medications are dosed together during the same treatment cycle, is expected to add extra logistical challenges, especially when it comes to biologics administered via different routes, such as, subcutaneously and intravenously.

From the above considerations, it becomes evident that logistics for in-home care need to be clearly defined as a guidance for local home health coordinators (Emiliano et al., 2017). This could involve nurses delivering the medication, or a dedicated professional care at-home service, specifically trained to provide physician-directed interprofessional care in patients' homes or other comparable settings (Vaartio-Rajalin & Fagerström, 2019). Measures, such as refrigeration and secure transportation boxes, need to be introduced to minimize potential damage to the product and thereby avoid harm to the patient and reduce potential errors. Delivery and transport under cold chain requirements have to be well organized to ensure efficacy and safety of the biologic medication (Laptoš & Omersel, 2018). Also, adequate and correctly sized sharps bins and safe drop-off points (e.g., general practitioner practices, health centers, or local pharmacies) or home pick-up services need to be organized upfront. Within this framework, so-called smart sharps bins that constantly monitor and track medication disposal represent an opportunity to measure adherence to treatment in a noncontrolled home environment (Gu et al., 2020).

Besides a demonstrated acceptable safety and tolerability profile of a medication for the individual patient, the precondition for allowing parenteral home- and self-administration is a product presentation that is straightforward and easy to use for a patient or lay caregiver. Such medications might even omit the need for a healthcare provider or nursing service to guide and support the dosing procedure. Namely, parenteral medications with minimal preparation needs that are administered with a fixed dose, or, in case of combination therapy with two or more biotherapeutics, as a fixed-dose combination (FDC), support a simple application procedure. Also, more convenient administration routes, such as subcutaneous dosing, would generally be preferred over more invasive and complex intravenous infusions. Ideally, dosing would be facilitated by means of an ergonomic and ready-to-use injection device. To account for different requirements, preferences, and capabilities of individual patients, more than one delivery option should be made available (Bittner & Schmidt, 2021).

Table 4.1 summarizes patient, healthcare provider, and treatment barriers to a flexible care setting and how to overcome them.

Home-administration of biologics in oncology—the regulatory perspective

The majority of biotherapeutics in oncology are not approved for administration in the home setting. To evaluate the regulatory requirements to avoid the restriction for use of the medication in a healthcare institution only, the following section details the related regulatory needs for blinatumomab (Blincyto), as an example for an intravenous biologic with a complex administration procedure, however, offering the option of receiving treatment at home. Moreover, the label languages for biologics approved for subcutaneous self-administration in the European Union in other disease areas will be summarized.

Regulatory precedent assessment—blinatumomab

The bi-specific antibody blinatumomab is used as a treatment for Philadelphia chromosome-negative relapsed or refractory acute lymphoblastic leukemia (Newman & Benani, 2016). Adverse events associated with blinatumomab treatment include cytokine release syndrome and neurotoxicity (Stein et al., 2019). The medication is available as a lyophilized powder for reconstitution in a single-use vial for continuous intravenous infusion. The comparatively complex dosing procedure requires aseptic handling by trained personnel and involves both a fixed dose as well as a body surface area–adjusted schedule for patients weighing less than 45 kg (Blincyto: highlights of prescribing information, 2021). Notwithstanding the complexity inherent to this dosing regimen, an outpatient administration option was considered "hugely beneficial to patients and would have a major impact on their quality of life" by the National Institute for Health and Care Excellence (NICE) in the United Kingdom on its blinatumomab appraisal (NICE – National Institute for Health & Care Excellence, 2017).

The Highlights of the Prescribing Information (United States) for blinatumomab do not specifically mention home-administration, but indirectly allow for dosing outside a hospital as follows: "Hospitalization is recommended for

Table 4.1 Patient, healthcare provider, and treatment barriers to a flexible care setting and how to overcome them.

	Identified barriers to flexible care setting	**Measures to overcome barriers**
Stakeholder		
Patient	Feeling uncomfortable with parenteral administration of biotherapeuticals without professional supervision	Establish strong communication network between patient, caregiver, and healthcare provider
	Feeling isolated without direct contact to healthcare provider or other patients	Treatment initiation by an experienced physician, proper training of eligible patients prior to dosing outside of healthcare institution
		Establish home healthcare services
		Telemedicine, i.e., adherence trackers, dosing reminders, remote dialog with oncologist
Healthcare provider	Same as for patients	Same as or patients
	Fearing loss of direct patient interactions, while still being liable for patient treatment	Expand value-based reimbursement (i.e., bundled payment models) to in-home healthcare
	Lacking reimbursement mechanisms to apply connected health services	
Treatment logistics and conduct	Complexity around parenteral administration	Ready-to-use subcutaneous dosing alternatives, including (connected) devices
	Patients' homes may not be set up to support logistics of parenteral administration	Clearly defined logistics for in-home care as a guidance for local home health coordinators
		Well organized transport under cold chain requirements
		Provision of refrigerators and secure transportation boxes
		Upfront organization of (smart) sharps bins and safe drop-off points or home pick-up services

the first 3 days of the first cycle and the first 2 days of the second cycle. For all subsequent cycle starts and re-initiations (e.g., if treatment is interrupted for 4 or more hours), supervision by a healthcare professional or hospitalization is recommended." (Blincyto: highlights of prescribing information, 2021). The Summary of Product Characteristics (SmPC; European Union) specifically mentions the home-administration option as follows: "Your doctor will discuss with you if you can continue treatment at home after your initial hospital stay. Treatment may include a bag change by a nurse." (Blincyto: summary of product characteristics, 2018).

According to the US Food and Drug Administration's (FDA) review documentation, 22 patients (4.6%) from the pooled patient population including 475 patients experienced a medication error in blinatumomab trials. The "causes for these events include: increased flow rate of the infusion pump due to pump malfunction as the most common error resulting in an overdose event, infusion pump rate set incorrectly, error in connecting the infusion line to the pump, and pharmacy preparation errors in calculating the correct concentration of the blinatumomab solution" (Blincyto: risk assessment & risk mitigation reviews, 2014). Both, the FDA and the European Medicines Agency (EMA) specifically pointed out the errors occurring either during the preparation of blinatumomab's solution for infusion or during the use of the infusion pump (Blincyto: risk assessment and risk mitigation review(s), 2014; European Public Assessment Report, 2021).

A number of risk mitigation requirements were consequently implemented. In the United States, a Risk Evaluation and Mitigation Strategy (REMS) was put in place. REMS is a drug safety program that the FDA can require for certain medications with serious safety concerns to help ensure the benefits of the medication outweigh its risks (Chan et al., 2018). REMS materials must be prominently displayed and disseminated at all relevant scientific meetings. REMS letters to healthcare providers, such as doctors, nurses, hospital and home healthcare pharmacists, and professional societies address the risk of preparation and administration errors associated with use of blinatumomab and are posted on the Blincyto REMS webpage (Blincyto: risk evaluation & mitigation strategy REMS, 2021).

The EMA requested the conduct of an observational postmarketing study to include frequency and types of blinatumomab medication errors as a primary goal and, additionally, healthcare providers, patients, and caregivers had to be provided with supporting educational material (Blincyto: summary of risk management plan, 2015).

Overall, in spite of the notable dosing complexity associated with intravenous infusion and a weight-dependent dose calculation step, for eligible patients and under the supervision or assistance of a healthcare provider, blinatumomab can be dosed in the home setting following treatment initiation in the clinic. Today, a number of biotherapeutics with subcutaneous dosing alternatives have entered the oncology market. Such offerings should provide a basis for reinforcing a flexible care setting in oncology as an option with increasing importance, especially in view of the COVID-19 pandemic.

Regulatory precedent assessment—home- and patient self-administration of biologics in other disease areas

In accessing possible product profiles for biologics envisaged for home- and patient self-administration in cancer care, biotherapeutics from other indications can serve as a precedent. Thus, to gain further insights into associated label languages, obligations, restrictions, and risk mitigation approaches, Table 4.2 highlights relevant sections in the SmPC for subcutaneously administered biotherapeutics with a self-administration label in different disease areas. The SmPC is a legal document approved as part of the marketing authorization of each medicine and the source of information for healthcare professionals on how to use the medication. Its information is updated throughout the lifecycle of the product as new data emerges (Summary of product characteristics – European Medicines Agency EMA, 2021).

Selected subcutaneous biotherapeutics include examples for both low- (up to approximately 2 mL) and high-volume (> approximately 2 mL) dosing solutions. Namely, the interleucin-1 receptor antagonist anakinra, the interleukin-6 receptor antagonists tocilizumab and sarilumab, the tumor necrosis factor-α inhibitors adalimumab, certolizumab pegol, etanercept, and golimumab, the selective costimulating modulator abatacept for the treatment of RA, as well interferons and the anti-CD20 mAb ofatumumab in MS represent volumes of up to 2 mL. Examples for high dosing volumes are immunoglobulins in primary immunodeficiency and evolocumab, a proprotein convertase subtilisin/kexin type 9 for the treatment of hyperlipidemia (Table 4.2).

Product presentation and dosing regimen

Product presentation

For the majority of self-administered products, dosing volumes do not exceed 1–2 mL. Typically, injection of these low-volume products is

Table 4.2 Excerpts from SmPC labels for biotherapeutics approved for patient self-administration (adult indications).

Molecule (brand name)	SC product presentations	Dosing regimen	Posology/self-administration *SmPC* Section 4.2 *"Posology and method of administration"*	Injection sites/recommendations	Tolerability (hypersensitivity/anaphylaxis/injection site reactions) *SmPC* Section 4.8 *"Undesirable effects"*
Interleucin-1 receptor antagonist					
Anakinra (Kineret)	- 0.67 mL in PFS	- q1d - Fixed dose regimen	- Treatment should be initiated and supervised by specialist physicians experienced in the diagnosis and treatment of RA - Cooling of the injection site, warming the injection liquid to room temperature, use of cold packs (before and after the injection), and use of topical glucocorticoids and antihistamines after the injection can alleviate the signs and symptoms of injection site reactions	- Abdomen (except for area around navel) - Top of thighs - Upper outer areas of buttocks - Outer area of upper arms - Alternating injection sites	- Anaphylaxis: uncommon - Injection site reactions: very common
Interleukin-6 receptor antagonists (RA indication)					
Sarilumab (Kevzara)	- 1.14 mL in PFS - 1.14 mL in PFP	- q2w - Fixed dose regimen (blood parameter-dependent dose adjustment)	- Treatment should be initiated and supervised by healthcare professionals experienced in the diagnosis and treatment of RA - A patient may self-inject Kevzara or the patient's caregiver may administer Kevzara if their healthcare professional determines that it is appropriate - Proper training should be provided to patients and/or caregivers on the preparation and administration of Kevzara prior to use	- Abdomen - Thigh - Upper arm - Rotating injection sites - No injection into skin that is tender, damaged, or has bruises or scars	- Hypersensitivity: has been reported (most common: injection site rash, rash, and urticaria) - Injection site erythema/pruritus: common

Tocilizumab (Actemra)	- 0.9 mL in PFS - 0.9 mL in PFP	- q1w and q2w - Fixed dose regimen (response- and blood parameter-dependent dose adjustment)	- Treatment should be initiated by healthcare professionals experienced in the diagnosis and treatment of RA - The first injection should be performed under the supervision of a qualified health care professional - A patient or parent/guardian can self-inject RoActemra only if the physician determines that it is appropriate and the patient or parent/guardian agrees to medical follow-up as necessary and has been trained in proper injection technique - All patients treated with RoActemra should be given the Patient Alert Card - Suitability of the patient or parent/guardian for subcutaneous home use should be assessed and patients or parent/guardian instructed to inform a healthcare professional before administering the next dose if they experience symptoms of an allergic reaction. Patients should seek immediate medical attention if developing symptoms of serious allergic reactions - After proper training in injection technique, patients may self-inject with RoActemra if their physician determines that it is appropriate	- Abdomen - Thigh - Upper arm - Rotating injection sites - No injection into moles, scars, or areas where the skin is tender, bruised, red, hard, or not intact	- Hypersensitivity: common - Anaphylaxis (fatal): rare - Injection site reactions: very common

(*Continued*)

Table 4.2 (Continued)

Molecule (brand name)	SC product presentations	Dosing regimen	Posology/self-administration *SmPC* Section 4.2 *"Posology and method of administration"*	Injection sites/recommendations	Tolerability (hypersensitivity/anaphylaxis/injection site reactions) *SmPC* Section 4.8 *"Undesirable effects"*
Tumor necrosis factor alpha antagonists (RA indication)					
Adalimumab (Humira)	- 0.8 mL vial - 0.2, 0.4, 0.8 mL in PFS - 0.4, 0.8 mL in PFP	- q2w - Fixed dose regimen	- Treatment should be initiated and supervised by specialist physicians experienced in the diagnosis and treatment of conditions for which Humira is indicated - Patients treated with Humira should be given the Patient Reminder Card - After proper training in injection technique, patients may self-inject with Humira if their physician determines that it is appropriate and with medical follow-up as necessary and with medical follow-up as necessary	- Front of the thighs - Abdomen (at least 5 cm from navel) - At least 3 cm from last injection site	- Hypersensitivity: common - Anaphylaxis: rare - Injection site reactions: very common
Certolizumab pegol (Cimzia)	- 1 mL in PFS - 1 mL in PFP - 1 mL in DDC	- q2w (starting dose), q2w or q4w (maintenance dose, response-dependent) - Fixed dose regimen	- Treatment should be initiated and supervised by specialist physicians experienced in the diagnosis and treatment of conditions for which Cimzia is indicated - Patients should be given the special reminder card - After proper training in injection technique, patients may self-inject using the PFS/PFP if their physician determines that it is appropriate and with medical follow-up as necessary	- Abdomen - Thigh - Rotating injection sites - No injection in an area where the skin is reddened, bruised, or hard	- Hypersensitivity (including anaphylactic shock): uncommon - Injection site reactions: common

			- The physician should discuss with the patient which injection presentation option is the most appropriate - The PFS with needle guard should only be used by healthcare professionals - Cimzia solution for injection in a DDC is intended for single use in conjunction with the electromechanical injection device named ava. After proper training in the injection technique, patients may self-inject using the electro mechanical injection device ava with the single-use DDC if their physician determines that it is appropriate and with medical follow-up as necessary - The physician should discuss with the patient which injection presentation option is the most appropriate		
Etanercept (Enbrel)	- Powder in vial and 1 mL solvent in PFS - 0.5, 1 mL in PFS - 0.5, 1 mL in PFP	- q1w or twice weekly - Fixed dose regimen	- Enbrel treatment should be initiated and supervised by specialist physicians experienced in the diagnosis and treatment of RA - Patients treated with Enbrel should be given the Patient Card - Enbrel powder for solution must be reconstituted in 1 mL of solvent before use	- Front of the middle thighs - Abdomen (except for the 5-cm area right around the navel) - Outer area of the upper arms - If self-injecting, outer area of the upper arms should not be used	- Serious anaphylactic reactions: rare - Injection site reactions: very common

(Continued)

Table 4.2 (Continued)

Molecule (brand name)	SC product presentations	Dosing regimen	Posology/self-administration *SmPC* Section 4.2 *"Posology and method of administration"*	Injection sites/recommendations	Tolerability (hypersensitivity/ anaphylaxis/injection site reactions) *SmPC* Section 4.8 *"Undesirable effects"*
Golimumab (Simponi)	- 0.5, 1 mL in PFS - 0.45 mL (P), 0.5, 1 mL in PFP	- q1m - Fixed dose regimen	- Treatment is to be initiated and supervised by qualified physicians experienced in the diagnosis and treatment of RA - Patients treated with Simponi should be given the Patient Reminder Card - After proper training in subcutaneous injection technique, patients may self-inject if their physician determines that this is appropriate, with medical follow-up as necessary	- Front of thighs (recommended) - Lower abdomen (except 5-cm area around belly-button) - Back of upper arms (caregiver only) - Rotating injection sites (different site within your preferred area for each injection) - No injection into skin that is tender, bruised, red, scaly, hard or has scars	- Hypersensitivity: common - Anaphylaxis: rare - Injection site reactions: common
Selective T cell costimulation modulator (RA indication)					
Abatacept (Orencia)	- 0.4, 0.7, 1 mL in PFS - 1 mL in PFP	- q1w - Fixed dose regimen (with/without IV loading dose)	- Treatment should be initiated and supervised by specialist physicians experienced in the diagnosis and treatment of RA - ORENCIA is intended for use under the guidance of a healthcare professional - After proper training in subcutaneous injection technique, a patient or caregiver may inject with ORENCIA if a physician/ healthcare professional determines that it is appropriate	- Abdomen - Front of the thighs - Outer area of upper arm (only if caregiver administered) - Rotation of injection sites - No injections into areas where the skin is tender, bruised, red, or hard	- Hypersensitivity: uncommon - Anaphylaxis: rare - Injection site reactions: common

Interferon					
Interferon beta-1a (Rebif)	- 0.2, 0.5 mL in PFS - 0.2, 0.5 mL in PFP - 1.5 mL in MDC	- Three times per week - Fixed dose regimen (up-titration scheme, tolerability-dependent)	- Treatment should be initiated under supervision of a physician experienced in the treatment of the disease - Prior to injection and for an additional 24 h after each injection, an antipyretic analgesic is advised to decrease flu-like symptoms associated with Rebif administration - PFP: RebiDose is a ready to use PFP for subcutaneous injection. It is intended for single use and should only be used following adequate training of the patient and/or carer - MDC: Rebif solution for subcutaneous injection in a cartridge is intended for multidose use with the RebiSmart electronic injection device following adequate training of the patient and/or carer	- Lower abdomen - Upper thighs - No injection into areas in which with lumps, firm knots, or pain	- Hypersensitivity (serious allergic reactions): rare - Anaphylaxis: rare - Injection site reactions: very common
Interferon beta-1b (Betaferon)	- Powder in vial and 1.2 mL solvent in PFS - AI (not specified)	- q2d - Fixed dose regimen (up-titration scheme, tolerability-dependent)	- The treatment with Betaferon should be initiated under the supervision of a physician experienced in the treatment of the disease	- Abdomen - Arm - Thigh - Buttocks - Rotating injection sites - No injection into an area with lumps, bumps, firm knots, pain or an area that is discolored, indented, scabbed, or where the skin is broken - Incidence of injection site reactions may be reduced by the use of an AI	- Hypersensitivity (serious): rare - Anaphylaxis: rare - Injection site reactions: frequently observed

(*Continued*)

Table 4.2 (Continued)

Molecule (brand name)	SC product presentations	Dosing regimen	Posology/self-administration *SmPC* Section 4.2 *"Posology and method of administration"*	Injection sites/recommendations	Tolerability (hypersensitivity/anaphylaxis/injection site reactions) *SmPC* Section 4.8 *"Undesirable effects"*
Interferon beta-1b (Extavia)	- Powder in vial and 1.2 mL solvent in PFS - AI (ExtaviPro 30 G)	- q2d - Fixed dose regimen (up-titration scheme, tolerability-dependent)	- The treatment with Extavia should be initiated under the supervision of a physician experienced in the treatment of the disease	- Abdomen - Arm - Thigh - Buttocks - Injection site rotation - Incidence of injection site reactions may be reduced by the use of an AI	- Hypersensitivity: potential risk - Anaphylaxis: rare - Injection site reactions: frequently
Peg-interferon beta-1a (Plegridy)	- 0.5 mL in PFS - 0.5 mL in PFP	- q2w - Fixed dose regimen (up-titration scheme)	- Treatment should be initiated under supervision of a physician experienced in the treatment of MS - Plegridy may be administered SC using a single-use PFP or PFS or IM using a single-use PFS - It is recommended that a healthcare professional trains patients in the proper technique for self-administering SC injections using the SC PFS/PFP or IM injections using the IM PFS as appropriate - Each Plegridy PFP/PFS for SC is provided with the needle preattached. Plegridy PFS for IM use is supplied as a PFS with a separate needle for IM use. Both IM and SC PFS and SC PFP are for single use only and should be discarded after use	- Abdomen - Arm - Thigh (SC and IM) - Rotation of injection sites - No direct injection into belly button - No injection into an area of the body where the skin is irritated, tender, red, bruised, tattooed, infected, or scarred	- Hypersensitivity: uncommon - Anaphylaxis: not known - Injection site reactions: common, very common (depending on type of reaction)

mAb against the protein CD20					
Ofatumumab (Kesimpta)	- 0.4 mL in PFS - 0.4 mL in PFP	- q1w (initial doses), q1m (maintenance doses) - Fixed dose regimen	- Treatment should be initiated by a physician experienced in the management of neurological conditions - Medicinal product is intended for patient self-administration - The first injection should be performed under the guidance of a healthcare professional	- Abdomen - Thigh - Upper outer arm	- Injection site reactions: very common
Immunoglobulin					
Human normal immunoglobulin (Hizentra)	- 5, 10, 20, 50 mL vial - 5, 10, 20 mL in PFS	- Dose and dosing regimen indication-dependent - BW-adjusted regimen (loading dose and response-dependent maintenance doses)	- Therapy should be initiated and monitored under the supervision of a healthcare professional experienced in the treatment of immunodeficiency/CIDP with SCIg - Subcutaneous infusion for home treatment must be initiated and monitored by a healthcare professional experienced in the guidance of patients for home treatment - The healthcare professional must select the appropriate way of infusion (device-assisted or manual push infusion), based on patient's individual medical situation and preferences - Infusion devices appropriate for subcutaneous administration of immunoglobulins can be used	- Abdomen - Thigh - Upper arm - Lateral hip - Each single dose may need to be injected at different anatomic sites - More than one infusion device can be used simultaneously - Up to 50 mL/site (adults) - No limit to number of infusion sites - Infusion sites should be at least 5 cm apart	- Hypersensitivity: rare - Anaphylaxis: unknown - Injection site reactions: very common

(Continued)

Table 4.2 (Continued)

Molecule (brand name)	SC product presentations	Dosing regimen	Posology/self-administration *SmPC* Section 4.2 *"Posology and method of administration"*	Injection sites/recommendations	Tolerability (hypersensitivity/ anaphylaxis/injection site reactions) *SmPC* Section 4.8 *"Undesirable effects"*
			- The patient or a caregiver must be instructed and trained in the use of infusion devices, the keeping of treatment diary, recognition of and measures to be taken in case of severe adverse reactions - Hizentra can be infused using an infusion device, or by manual push with a syringe. The recommended initial infusion rate depends on the individual patient's needs		
Human normal immunoglobulin and recombinant human hyaluronidase (Hyqvia)	- 25, 50, 100, 200, 300 mL in vial (SCIg) - 1.25, 2.5, 5, 10, 15 mL in vial (rHuPH20)	- Dose and dosing regimen indication-, PK- and response-dependent - BW-adjusted regimen (loading dose and response-dependent maintenance doses)	- Replacement therapy should be initiated and monitored under the supervision of a physician experienced in the treatment of immunodeficiency - In case subcutaneous infusion of HyQvia is used for home treatment, therapy should be initiated and monitored by a physician experienced in the guidance of patients for home treatment. The patient will be instructed in infusion techniques, the use of an infusion pump or syringe driver, the keeping of a treatment diary, recognition of possible severe adverse reactions and measures to be taken in case these occur	- Middle to upper abdomen - Thigh - Rotating injection sites - If two sites are used, the two infusion sites should be on contra lateral sides of the body - Avoidance of bony areas, visible blood vessels, scars and any areas of inflammation or infection	- Hypersensitivity: rare - Anaphylaxis: rare - Injection site reactions: very common, common, uncommon (depending on type of reaction)

			- HyQvia can be used to administer a full therapeutic dose in one to two sites up to every 4 weeks - The IG 10% component should be infused using a pump. The rHuPH20 may be hand-pushed or infused by a pump - It is recommended that the rHuPH20 component be administered at a constant rate and that the rate of administration of the IG 10% should not be increased above the recommended rates, particularly when the patient has just started with HyQvia therapy - First, the full dose of rHuPH20 solution is infused at a rate of 1–2 mL/min per infusion site or as tolerated. Infuse the full dose per site of IG 10% through the same subcutaneous needle set within 10 min of the rHuPH20		
Proprotein convertase subtilisin/kexin type 9 inhibitor					
Evolocumab (Repatha)	- 1 mL in PFS - 1 mL in PFP (SureClick) - 3.5 mL in C (AMD)	- q2w, q1m Fixed dose regimen (indication-dependent up-titration, response-dependent)	- Repatha is intended for patient self-administration after proper training. Administration of evolocumab can also be performed by an individual who has been trained to administer the product	- Abdomen (except for 5 cm around the belly button) - Thigh - Upper arm region (caregiver only) - Rotation of injection sites - No injection into areas where the skin is tender, bruised, red, or hard	- Hypersensitivity: common - Injection site reactions: common

AI, Autoinjector; *AMD*, automated mini-doser; *BW*, body weight; *CIPD*, chronic inflammatory demyelinating polyneuropathy; *C*, cartridge; *cm*, centimeters; *DDC*, dose-dispenser cartridge; *IG 10%*, human normal immunoglobulin 10%; *MDC*, multidose cartridge; *mL*, milliliters; *IM*, intramuscular; *IV*, intravenous; *MS*, multiple sclerosis; *P*, pediatric; *PFS*, prefilled syringe; *PFP*, prefilled pen; *PK*, pharmacokinetic; *q1w*, weekly; *q2w*, every 2 weeks; *q1m*, monthly; *RA*, rheumatoid arthritis; *rHuPH20*, recombinant human hyaluronidase; *SC*, subcutaneous; *SCIg*, human normal immunoglobulin.

facilitated by means of ready-to-use PFSs, PFPs or AIs. In addition, some products are presented as a powder for injection and associated solvent for dilution, thus, requiring a reconstitution step prior to injection, conducted by either a healthcare professional, the patient, or a lay caregiver.

For example, the interferon beta-1b presentation [Extavia (EMC – SmPC Database, 2021)] is a powder for reconstitution in a vial which in a first step is diluted using 1.2 mL of solvent in a PFS with a needle or a vial adapter. After reconstitution, 1 mL of the solution is drawn from the vial into the syringe for the actual subcutaneous injection. In addition, the SmPC refers to the use of an AI (ExtaviPro 30 G autoinjector) and the instructions for use provided with this device. The AI has been reported to significantly improve overall patient satisfaction, including satisfaction associated with effectiveness, side effects, and convenience in MS patients (Hoffmann et al., 2017).

Also etanercept [Enbrel (EMC – SmPC Database, 2021)] is available as a powder for reconstitution, in addition to PFS and PFP presentations. The powder is reconstituted with 1 mL water for injections before subcutaneous administration. The preparation and dosing procedures and corresponding equipment are comparable to those of Extavia. The etanercept formulation does not contain an antibacterial preservative, and therefore, solutions prepared with water for injections should be administered as soon as possible and within 6 hours following reconstitution [Enbrel (EMC – SmPC Database, 2021)].

Certolizumab pegol's product presentations include a solution in a disposable dose-dispenser cartridge [Cimzia (EMC – SmPC Database, 2021)], that is intended for single-use in conjunction with the reusable electromechanical self-injection device Ava that was developed in close cooperation with patients (Pouls et al., 2020). The device includes the option to personalize administration in order to help overcome self-injection challenges such as dexterity problems, uncertainty during self-injection, and forgetfulness. The device also includes an injection log that may facilitate a dialog about treatment adherence between patient and healthcare professional (Tatla et al., 2020).

High-volume subcutaneous products for self-administrations are presented in large size vials and PFSs. The human normal immunoglobulin Hizentra (EMC – SmPC Database, 2021), for example, is available in different PFS and vial presentations with volumes of up to 20 and 50 mL, respectively. Volumes of up to 50 mL are injected either by manual push using a syringe and butterfly needle or with the help of an infusion pump.

The manual push methodology allows the patient to inject at the most comfortable rate, while pump-assisted infusions generally allow infusions of larger volumes and thus require fewer needle sticks and less frequent infusions (Cowan et al., 2021).

The product presentations for the human normal immunoglobulin Hyqvia (EMC – SmPC Database, 2021) consist of vials with volumes of up to 300 mL containing the immunoglobulin and vials with volumes of up to 15 mL containing recombinant human hyaluronidase. Hyaluronidase is an enzyme that temporarily breaks down hyaluronan, a component of the extracellular matrix and, thus, eases subcutaneous administration of larger volumes of immunoglobulin at a single injection site (Wasserman, 2017). Consequently, the immunoglobulin only needs to be administered every 3–4 weeks. Hyaluronidase is preinfused subcutaneously to allow cleavage of hyaluronan, followed by volumes of up to 600 mL of the immunoglobulin per injection site. Dosing can be further facilitated by means of an infusion pump.

Dosing regimen

Beside the product presentation another aspect that impacts the ease of administering a medicine is the dosing regimen. For biologics, the most convenient dosing schedule would be based on a fixed dose with the same dose level and dosing frequency for all patients in a given indication. Drug administration can be complicated with body weight- or body surface area-adjusted dosing, different loading and maintenance doses, tolerability-based up-titration schemes, as well as efficacy- and safety-dependent dose and frequency changes. As such, complex dosing regimens can result in individual dose calculation errors, missing doses, and related drug wastage (Novakovic et al., 2019), and self- or caregiver-assisted administration without the supervision of a healthcare provider can be challenging.

In this context, notably, a number of the subcutaneous self-administration examples require dose calculation and dose adaptation throughout the treatment period. All biotherapeutics for the treatment of adults with RA are approved with a body size–independent fixed dose, so no dose calculation step is required. Sarilumab is administered with a fixed dose and dosing regimen, but the total dose may have to be reduced depending on the individual patient's blood parameters [Kevzara (EMC – SmPC Database, 2021)]. Equally, in RA, subcutaneous tocilizumab is dosed independent of body size, and in monotherapy, some patients who experience a decrease in their response to the standard dose and dosing regimen may receive the same dose with a more frequent regimen or twice the dose while maintaining the original dosing frequency [Actemra (EMC – SmPC Database, 2021)]. The certolizumab pegol dosing regimen starts with a fixed loading dose for three treatment cycles, which is followed by lower maintenance doses with the same dosing regimen. Once the clinical response is confirmed, patients may receive a higher dose with a less frequent dosing regimen [Cimzia (EMC – SmPC Database, 2021)]. Treatment with abatacept may be initiated with an intravenous loading dose followed by regular subcutaneous fixed doses [Orencia (EMC – SmPC Database, 2021)]. A loading dose is typically administered with the aim to achieve steady-state serum drug concentrations earlier after treatment starts (Geurts-Voerman et al., 2020).

The interferons [Rebif, Extavia (EMC – SmPC Database, 2021)] and the pegylated interferon [Plegridy (EMC – SmPC Database, 2021)] for the treatment of MS are all administered with a fixed dose and it is recommended to initiate treatment with an up-titration scheme to reduce adverse reactions [Rebif, Extavia, Plegridy (EMC – SmPC Database, 2021)].

For the high-volume subcutaneous self-administration products, dosing regimens of human normal immunoglobulins are indication- and response-dependent, thus requiring individual dose adaptation [Hizentra, Hyqvia (EMC – SmPC Database, 2021)]. The lipid lowering mAb evolocumab is administered with fixed doses and different dosing regimens. Depending on the product presentation, the higher dose is either given with three PFSs or PFPs administered consecutively within 30 minutes or with one higher volume on-body delivery system (OBDS) [Repatha (EMC – SmPC Database, 2021)].

Safety and tolerability profile

In order to decide on whether a subcutaneously dosed biological medicine is eligible for administration outside of a healthcare institution, safety and tolerability information from clinical trials has to be taken into consideration. Key focus in this context is on hypersensitivity reactions predominantly occurring during the initial treatment cycles. Typical symptoms vary in severity from mild to severe and include flushing, chills, fever, tachycardia, hypertension, dyspnea, nausea, vomiting, and syncope (Isabwe et al., 2018). These reactions may however evolve into anaphylaxis, generalized as a serious life-threatening or systemic hypersensitivity reaction (Turner et al., 2019), requiring accessibility of resuscitation facilities in close proximity to the healthcare facility (Bateman & Ferner, 2006). Consequently, subcutaneous treatments with biotherapeutics will usually be initiated in the clinic and depending on their individual tolerability profile, appropriately trained patients may be allowed to continue dosing outside of a controlled setting.

As per Section 4.8 "Undesirable effects" of the SmPC and across the selected self-administration examples, hypersensitivity reactions have been observed, and in the majority of cases, anaphylactic reactions are either rare (incidence $<1/10{,}000$) or not reported. Other frequently observed events include injection site reactions, localized skin reactions that occur at the injection site following subcutaneous administration. Such reactions are typically associated with erythema, edema, itch, and/or pain (Grace et al., 2019). For the selected examples, injection site reactions are observed very commonly (incidence $\geq 1/10$) or commonly (incidence $\geq 1/100$ to $<1/10$) in clinical trials, but do not generally prevent self-administration at home (Summary of product characteristics – EMA, 2021).

Dosing instructions

Instructions on treatment initiation and conduct are given in Section 4.2 "Posology and method of administration" of the SmPC. In the following, this section is used as a basis for the assessment of general dosing and administration instructions for self-administered biotherapeutics.

Across all selected examples, treatment should be initiated and supervised by a physician experienced in the management of patients in the given indication [Kineret, Kevzara, Actemra, Humira, Cimzia, Enbrel, Simponi, Orencia, Rebif, Betaferon, Extavia, Plegridy, Kesimpta, Hizentra, Hyqvia (EMC – SmPC Database, 2021)]. This approach ensures that patient eligibility for self-administration can be judged by the responsible healthcare provider who will be in a position to arrange and conduct adequate training for the patient and possibly a lay caregiver. Some SmPCs specify that self- or caregiver-assisted administration is feasible, if a healthcare professional determines that it is appropriate [Kevzara, Actemra, Humira, Cimzia, Simponi, Orecia (EMC – SmPC Database, 2021)] and following proper training in the injection technique has been conducted [Kineret, Kevzara, Actemra, Humira, Cimzia, Simponi, Orencia, Plegridy, Hyqvia, Repatha (EMC – SmPC Database, 2021)]. As part of the training procedures, patients are given advice about which injection presentation is the most appropriate based on the patient's individual medical situation and preferences [Cimzia, Hizentra (EMC – SmPC Database, 2021)]. For more detailed instructions, SmPCs principally refer to the package leaflet and the section on instructions for use. Some SmPCs provide a comprehensive description on how the medication has to be administered including measures on how to act in the event of known side effects in patient alert or reminder cards in Section 4.2 "Posology and method of administration" [Actemra, Humira, Cimzia, Enbrel, Simponi (EMC – SmPC Database, 2021)]. In addition, as described in the SmPCs for subcutaneous administration of high-volume immunoglobulins, the patient can be reminded to keep a treatment diary [Plegridy, Hyqvia (EMC – SmPC Database, 2021)], allowing the healthcare provider to retrospectively assess successful compliance with the dosing instructions.

A number of SmPCs list particular instructions in Section 4.2 "Posology and method of administration", aiming to ensure safe and convenient subcutaneous administration. For example, the interferon beta-1a SmPC specifically specifies that "prior to injection and for an additional 24 hours after each injection, an antipyretic analgesic is advised to decrease flu-like symptoms associated with Rebif administration" [Rebif (EMC – SmPC Database, 2021)]. Along the same lines, the anakinra SmPC recommends using cold packs before and after the injection and topical glucocorticoids and antihistamines to alleviate signs and symptoms of injection site reactions [Kineret (EMC – SmPC Database, 2021)].

Certolizumab pegol can be self-administered with a PFS, PFP, and dose dispenser cartridge, but is not suitable for self-administration when using the PFS with needle guard that should only be used by healthcare professionals [Cimzia (EMC – SmPC Database, 2021)]. Peginterferon beta-1a [Plegridy (EMC – SmPC Database, 2021)] can be self-administered both subcutaneously and intramuscularly. Here, Section 4.2 of the SmPC specifies that the PFP can only be used for subcutaneous administration. The PFS is suitable for both administration routes. For subcutaneous dosing, the PFS is provided with the needle preattached, and for intramuscular use, the product is supplied as a PFS with a separate needle.

Even more detailed dosing and administration instructions are provided for high-volume subcutaneous administrations of the human immunoglobulins Hizentra and Hyqvia. Hizentra can be infused using either an infusion device or by manual push with a syringe. The injection rates should be adjusted according to the individual patient's needs. For higher infusion rates and the manual push methodology, the use of a larger needle gauge is recommended [Hizentra (EMC – SmPC Database, 2021)]. The SmPC also frames that the injection rate differs depending on the injection method and individual treatment tolerability. Namely, if "the patient tolerates the initial infusions at the full dose per site and maximum rate, an increase in the infusion rate of successive infusions may be considered at the discretion of the patient and based on the healthcare professionals' judgement." Inherent to Hyqvia's (EMC – SmPC Database, 2021) more complex administration procedure, SmPC Section 4.2 details the injection rates and recommends needle sizes for the preinjected recombinant human hyaluronidase and the subsequently infused immunoglobulin solution.

Injection sites

The low-volume products (i.e., up to 2 mL injection volume) are approved for injection into either the thigh or the abdomen [Kineret, Kevzara, Actemra, Humira, Cimzia, Enbrel, Simponi, Orencia, Rebif, Betaferon, Extavia, Plegridy, Hizentra, Hyqvia, Repatha (EMC – SmPC Database, 2021)]; and the majority of products can also be injected into the arm [Kineret, Kevzara, Actemra, Enbrel, Simponi, Orencia, Betaferon, Extavia, Plegridy, Hizentra, Repatha, Kesimpta (EMC – SmPC Database, 2021)]. In addition, the buttocks [Kineret, Betaferon (EMC – SmPC Database, 2021)] and the hip [Hizentra (EMC – SmPC Database, 2021)] represent suitable injection sites, as recommended in several SmPCs.

Some SmPCs specify specific regions of the respective injection site (i.e., upper thighs, front of the middle thighs, upper outer areas of the buttocks, lower abdomen, middle to upper abdomen, back of upper arm, outer area of the upper arm, upper arm region, lateral hip). Frequently, the SmPCs recommend not to inject into the abdominal area around the navel [Kineret, Humira, Enbrel, Simponi, Plegridy, Repatha (EMC – SmPC Database, 2021)]. In some cases, injection into the arm has to be conducted by a caregiver, and this injection site is specifically not meant for self-administration [Enbrel, Orencia, Repatha (EMC – SmPC Database, 2021)]. In addition, SmPCs instruct rotating the injection site [Kineret, Kevzara, Actemra, Cimzia, Simponi, Orencia, Betaferon, Extavia, Plegridy, Hizentra, Hyqvia, Repatha (EMC – SmPC Database, 2021)] or alternatively to slightly move the injection area by a few centimeters while using the same injection site [Humira (EMC – SmPC Database, 2021)]. A number of SmPCs provide guidance on skin conditions that preclude using an injection site as follows: areas where the skin is not intact, scaly, tender, bruised, tattooed, infected, inflamed, scarred,

red, hard, discolored, indented, scabbed, broken or into areas with lumps, bumps, visible blood vessels, firm knots or pain [Kineret, Kevzara, Actemra, Cimzia, Simponi, Rebif, Betaferon, Hizentra, Hyqvia (EMC – SmPC Database, 2021)].

Subcutaneous administration of biotherapeutics in cancer care—a status update

Today, a number of subcutaneous formulations for mAbs have been approved in different oncology indications (refer to Chapter 1, Enabling customer-centric and sustainable drug delivery—formulation and device lifecycle management of biotherapeutics). Namely, high-volume subcutaneous formulations for trastuzumab (5 mL), rituximab (11.7 and 13.4 mL), daratumumab (15 mL), and a FDC of trastuzumab and pertuzumab (10 mL) are available as subcutaneous dosing alternatives to established intravenous administration. For lower dose subcutaneous treatments with injection volumes of up to approximately 2 mL established PFS, AI and PFP device platforms could principally also be used for cancer treatments. Device platforms for higher dosing volumes, that is, in the range of the above mAb formulations, are still in early development.

To assess the dosing complexity as one of the key elements when considering dosing in a flexible care setting, Table 4.3 compares the relevant parts in the SmPCs for the intravenous to the subcutaneous versions of the same mAbs. Focus is on Section 4.2 "Posology and method of administration." For all of these mAbs in oncology, dosing needs to be initiated and supervised by a physician. For both, the intravenous and subcutaneous presentations of pertuzumab (in combination with trastuzumab) and rituximab, dosing has to be conducted in an environment where full resuscitation facilities are immediately available (Summary of product characteristics – European Medicines Agency EMA, 2021). The SmPC specification is different for daratumumab (EMC – SmPC Database, 2021) that exhibits a reduced incidence and severity in infusion-related reactions following subcutaneous as compared to intravenous administration (Mateos et al., 2020). Here, intravenous infusion in general demands an environment where resuscitation facilities are available, while for subcutaneous injections, such facilities are only mandated for the first dose.

Trastuzumab for the treatment of human epidermal growth factor receptor 2 (HER2)-positive cancer types was the first high volume mAb available with a subcutaneous dosing alternative. Intravenous dosing is comparatively complex, as the product presentation is a lyophilisate for

Table 4.3 Excerpts from SmPC labels for high-volume mAbs approved for intravenous and subcutaneous administration in oncology.

Molecule (brand name)	Route	Product presentations	Dosing frequency/indications	Posology/self-administration *SmPC* Section 4.2 ***"Posology and method of administration"***	SC injection sites/recommendations
mAbs against HER2+ cancer					
Trastuzumab (Herceptin)	IV	- 150 mg powder for reconstitution in vial	- q1w, q3w - BW-adjusted dose regimen (loading and maintenance dose) - HER2 + mBC, eBC and mGC	- Herceptin loading dose should be administered as a 90-min intravenous infusion - Herceptin intravenous infusion should be administered by a healthcare provider prepared to manage anaphylaxis and an emergency kit should be available - Patients should be observed for at least 6 h after the start of the first infusion and for 2 h after the start of the subsequent infusions for symptoms like fever and chills or other infusion-related symptoms - Interruption or slowing the rate of the infusion may help control such symptoms. The infusion may be resumed when symptoms abate. If the initial loading dose was well tolerated, the subsequent doses can be administered as a 30-min infusion	NA

(Continued)

Table 4.3 (Continued)

Molecule (brand name)	Route	Product presentations	Dosing frequency/indications	Posology/self-administration *SmPC* Section 4.2 *"Posology and method of administration"*	SC injection sites/recommendations
	SC	- 600 mg/5 mL solution in vial	- q3w - Fixed dose regimen (no loading dose) - HER2 + mBC and eBC	- The 600 mg dose should be administered as a subcutaneous injection only over 2–5 min every 3 weeks - Patients should be observed for 6 h after the first injection and for 2 h after subsequent injections for signs or symptoms of administration-related reactions	- Thigh - Injection site should be alternated between the left and right thigh (at least 2.5 cm from the old site) - No injection into areas where the skin is red, bruised, tender, or hard - Other medicinal products for subcutaneous administration should preferably be injected at different sites
Pertuzumab (Perjeta)	IV	- 420 mg/ 14 mL concentrate for solution for infusion in vial	- q3w - Fixed dose regimen (loading and maintenance dose) Administered in combination with IV or SC trastuzumab - HER2 + mBC and eBC	- Perjeta should only be initiated under the supervision of a physician experienced in the administration of anticancer agents. Perjeta should be administered by a healthcare professional prepared to manage anaphylaxis and in an environment where full resuscitation facilities are immediately available - The recommended initial loading dose of pertuzumab is 840 mg	NA

				administered as a 60 min intravenous infusion, followed every 3 weeks thereafter by a maintenance dose of 420 mg administered over a period of 30–60 min - An observation period of 30–60 min is recommended after completion of each infusion. The observation period should be completed prior to any subsequent infusion of trastuzumab or chemotherapy - Perjeta and trastuzumab should be administered sequentially and not mixed in the same infusion bag. Perjeta and trastuzumab can be given in any order - When administered with Perjeta the recommendation is to follow a 3 weekly schedule for trastuzumab administered as either: • an IV infusion with an initial loading dose of trastuzumab 8 mg/kg body weight followed every 3 weeks thereafter by a maintenance dose of 6 mg/kg body weight or • a fixed subcutaneous dose of trastuzumab by injection (600 mg) every 3 weeks irrespective of the patient's body weight	

(*Continued*)

Table 4.3 (Continued)

Molecule (brand name)	Route	Product presentations	Dosing frequency/indications	Posology/self-administration *SmPC* Section 4.2 *"Posology and method of administration"*	SC injection sites/recommendations
				- For the initial dose, the recommended infusion period is 60 min. If the first infusion is well tolerated, subsequent infusions may be administered over a period of 30–60 min	
Trastuzumab + pertuzumab (Phesgo)	SC	- 1200 mg/600 mg/15 mL solution in vial - 600 mg/600 mg/10 mL solution in vial	- q3w - Fixed dose regimen (loading and maintenance dose) - HER2 + mBC and eBC	- Phesgo should only be initiated under the supervision of a physician experienced in the administration of anticancer agents. Phesgo should be administered by a healthcare professional prepared to manage anaphylaxis and in an environment where full resuscitation facilities are immediately available - The loading dose and maintenance dose should be administered over 8 and 5 min, respectively - An observation period of 30 min after completion of Phesgo loading dose and 15 min after completion of the maintenance dose is recommended for injection-related reactions	- Thigh - Injection site should be alternated between the left and right thigh (at least 2.5 cm from the old site) - Dose should not be split between two syringes or between two sites of administration - No injection into areas into areas where the skin is red, bruised, tender, or hard - Other medicinal products for subcutaneous administration should preferably be injected at different sites

mAb against the protein CD20					
Rituximab (MabThera)	IV	- 100 mg/10 mL concentrate for solution in vial - 500 mg/50 mL concentrate for solution in vial	- q1w, q2w, q3w, q4w, q2m, q3m, q6m - BSA-adjusted dose regimen (indication-dependent dose levels; induction and maintenance doses) - NHL, CLL, RA, PV, GPA, MPA	- MabThera should be administered under the close supervision of an experienced healthcare professional, and in an environment where full resuscitation facilities are immediately available - Premedication consisting of an antipyretic and an antihistaminic, e.g., paracetamol and diphenhydramine, should always be given before each administration of MabThera - The recommended initial rate for infusion is 50 mg/h; after the first 30 min, it can be escalated in 50 mg/h increments every 30 min, to a maximum of 400 mg/h - Subsequent doses of MabThera can be infused at an initial rate of 100 mg/h, and increased by 100 mg/h increments at 30 min intervals, to a maximum of 400 mg/h	NA

(Continued)

Table 4.3 (Continued)

Molecule (brand name)	Route	Product presentations	Dosing frequency/indications	Posology/self-administration *SmPC* Section 4.2 *"Posology and method of administration"*	SC injection sites/recommendations
	SC	- 1400 mg/ 11.7 mL solution in vial - 1600 mg/ 13.4 mL solution in vial	- q3w, q4w, q2m, q3m - Fixed dose regimens (BSA-adjusted IV loading doses) - NHL, CLL	- MabThera should be administered under the close supervision of an experienced healthcare professional, and in an environment where full resuscitation facilities are immediately available - Premedication consisting of an antipyretic and an antihistaminic, e.g., paracetamol and diphenhydramine, should always be given before each administration of MabThera - The 1400 mg strength is intended for subcutaneous use in NHL only - The 1600 mg strength is intended for subcutaneous use in CLL only - MabThera 1400 mg/1600 mg subcutaneous formulation should be administered as subcutaneous injection only, over approximately 5/7 min	- Abdomen - No injection into areas where the skin is red, bruised, tender, hard or areas where there are moles or scars - No data are available on performing the injection in other sites of the body, therefore injections should be restricted to the abdominal wall - Other medicinal products for subcutaneous administration should preferably be injected at different sites

mAb against the protein CD38					
Daratumumab (Darzalex)	IV	- 100 mg/5 mL concentrate for solution in vial - 400 mg/20 mL concentrate for solution in vial	- q1w, q2w, q3w, q4w - BW-adjusted dose regimen (3-step increase in dosing intervals; indication-dependent) - MM (different forms)	- DARZALEX should be administered by a healthcare professional, in an environment where resuscitation facilities are available - Pre- and postinfusion medicinal products should be administered to reduce the risk of infusion-related reactioSarilumabns with daratumumab	
	SC	- 1800 mg/15 mL solution in vial	- q1w, q2w, q3w, q4w - Fixed dose regimen (3-step increase in dosing intervals; indication-dependent) - MM (different forms); same as IV	- DARZALEX should be administered by a healthcare professional, and the first dose should be administered in an environment where resuscitation facilities are available - Pre- and postinjection medicinal products should be administered to reduce the risk of infusion-related reactions with daratumumab	- Abdomen (approximately 7.5 cm to the right or left of the navel over approximately 3–5 min) - Rotation of injection sites - No injection into areas where the skin is red, bruised, tender, hard or areas where there are scars - No data are available on performing the injection in other sites of the body, therefore injections should be restricted to the abdomen - Other medicinal products for subcutaneous administration should preferably be injected at different sites

BSA, Body surface area; *BW*, body weight; *CLL*, chronic lymphocytic leukemia; *eBC*, early breast cancer; *GPA*, granulomatosis with polyangiitis; *IV*, intravenous; *mBC*, metastatic breast cancer; *mGC*, metastatic gastric cancer; *MM*, multiple myeloma; *MPA*, microscopic polyangiitis; *NHL*, non-Hodgkin's lymphoma; *PV*, pemphigus vulgaris; *q1w*, weekly; *q2w*, every 2 weeks; *q3w*, every 3 weeks; *q4w*, every 4 weeks; *q2m*, every 2 months; *q3m*, every 3 months; *q6m*, every 6 months; *RA*, rheumatoid arthritis; *SC*, subcutaneous.

reconstitution. Intravenous trastuzumab is administered with a weight-adjusted dosing regimen with two alternative dosing frequencies and requires two different loading and maintenance dose levels [Trastuzumab (EMC – SmPC Database, 2021)]. To manage possible hypersensitivity or anaphylactic reactions, the loading dose is to be infused over a period of 90 minutes and, provided the initial infusion was well tolerated, all subsequent maintenance doses can be infused over 30 minutes.

With the subcutaneous version of trastuzumab, dosing complexity is markedly reduced. The subcutaneous dosing regimen is administered with the same fixed dose for all treatment cycles and only one dosing regimen with an injection duration of only 2–5 minutes. The formulation is a solution for infusion that can directly be injected by a healthcare provider using a handheld syringe (HHS). The solution is to be administered into the thigh only, as no other injection site was studied in the supporting clinical trials (Ismael et al., 2012; Wynne et al., 2013). Independent of the administration route, patients need to be observed for 6 hours after the first and for 2 hours after the subsequent injections, a condition that currently hinders nonsupervised home-administration.

A progression in subcutaneous administration in HER2-positive breast cancer is Phesgo. Phesgo is the first approved FDC of 2 mAbs. Namely, trastuzumab and pertuzumab are coformulated in the same 10 mL dosing vehicle. The intravenous counterparts are separate formulations of trastuzumab [Herceptin (EMC – SmPC Database, 2021)] and pertuzumab [Perjeta (EMC – SmPC Database, 2021)] that are sequentially infused, but not mixed in the same infusion bag. In line with a general evolution from body size–adjusted to fixed dosing, if permitted by the efficacy and tolerability profile of a molecule (Wang et al., 2009), intravenous pertuzumab is already available with fixed loading and maintenance doses and one dosing regimen (Kirschbrown et al., 2019). The intravenous pertuzumab loading dose is administered over 60 minutes and the subsequent maintenance doses over 30–60 minutes. The postinfusion observation period is 30–60 minutes and within this time period no trastuzumab or chemotherapy should be administered. Subsequent trastuzumab infusion can either be done with the intravenous or with the subcutaneous formulations. In combination with pertuzumab, the same observation times of 60 minutes for the first and 30–60 minutes for all subsequent infusions also apply for trastuzumab, independent of the administration route.

The liquid subcutaneous FDC of trastuzumab and pertuzumab [Phesgo (EMC – SmPC Database, 2021)] is based on a fixed loading dose and fixed maintenance doses with one dosing regimen only.

Compared to intravenous coadministration of pertuzumab and trastuzumab, the injection duration of the respective 15 and 10 mL dosing solutions is reduced to only 8 and 5 minutes and the observation periods for the first and subsequent doses are reduced to 30 and 15 minutes. The only permitted injection site is the thigh, as studied in the pivotal trials supporting registration of the subcutaneous formulation (Kirschbrown et al., 2019; Tan et al., 2021).

The intravenous formulation for rituximab, directed against the protein CD20, which is primarily found on the surface of immune system B cells, is either available as a 10 or 50 mL vial containing a liquid concentrate for solution for infusion. Depending on the indication, a number of dosing schedules, including induction and maintenance regimens, are approved, and the dose needs to be calculated per body surface-area for each individual patient [MabThera (EMC – SmPC Database, 2021)]. SmPC Section 4.2 "Posology and method of administration" recommends premedication consisting of an antipyretic and an antihistaminic before each administration of rituximab. The SmPC specifies different infusion rates for the first and subsequent dose levels that can be increased at 30-minute intervals from 50 mg/hour up to a predefined maximum rate of 400 mg/hour.

To account for different indications, subcutaneous rituximab is presented in two vial strengths containing 11.7 and 13.4 mL as solution for injection. The solution is administered manually by a healthcare provider using a HHS without further dilution. Different to subcutaneous trastuzumab or the subcutaneous pertuzumab/trastuzumab combination, the only permitted injection site is the abdomen, as studied in the pivotal clinical trials (Assouline et al., 2015; Davies et al., 2014; Davies et al., 2017). Injection time for the two dose strengths is approximately 5 and 7 minutes and the recommendation for premedication is the same as for the intravenous version.

Analog to rituximab, two vial sizes with a liquid concentrate for solution for infusion with 5 and 20 mL fill volume are available for the intravenous formulation of daratumumab, a mAb directed against CD38, overexpressed in multiple myeloma cells. Different dosing frequencies are approved to account for different patient populations. The dose is calculated on a weight-adjusted basis and there is an indication-dependent 3-step increase in dosing intervals [Darzalex (EMC – SmPC Database, 2021)].

The daratumumab subcutaneous solution in a vial is administered as a fixed dose with the same dosing frequencies as the intravenous equivalent.

For both administration routes, it is recommended to apply pre- and post-infusion medicinal products to account for infusion-related reactions. Daratumumab is to be administered in the abdomen only, as studied in supporting clinical trials (Usmani et al., 2019).

Evidently, all of these high-volume subcutaneous treatments are still dosed in a controlled setting and by a healthcare provider. To realize reliable and convenient dosing in a flexible care setting for eligible patients or by lay caregivers, a ready-to-use injection device would be mandated. Another drawback with the administration procedure that may compromise acceptance for self-administration in a broader patient population is that all of the above products are only approved for injection at a single anatomical site, the abdomen versus the thigh. Patients and healthcare providers prefer different injection sites (Ciftci & Avsar, 2016), and some may not adequately comply with home- or especially self-administration, if their favored site is not allowed. Looking at the SmPCs for the subcutaneous self-administration examples from other disease areas, typically both the thigh and the abdomen, and frequently also the arm are approved injection sites. Consequently, to further improve patient experience, the availability of more than one injection site should be explored further.

A key focus of future development programs for subcutaneous formulations will be on the adverse event profile compared to intravenous infusion. Subcutaneous daratumumab exhibits an overall adverse event profile comparable to intravenous administration, but a lower incidence and severity of infusion-related reactions, which mainly occurred with the first administration and mostly developed within 6 hours after the start of the infusion. It is hypothesized that the gradual absorption of daratumumab into systemic circulation after subcutaneous administration could be the underlying reason for this finding (Usmani et al., 2019). If this observation is also made for other mAbs in the future, and assuming an otherwise acceptable safety profile, subcutaneous dosing may indeed be an entry ticket for at-home or even patient self-administration.

Summary and outlook

In summary, crucial prerequisites for successfully shifting parenteral administration of suitable biologics in cancer care to the home setting and at the same time ensuring adherence and compliance with the dosing regimen are that (1) the patient acknowledges and understands the drug administration procedure and is capable of administering the medication independently or with the help of a caregiver, (2) oncologists, nurses, and supporting home care service providers are appropriately trained, equipped, and reimbursed for providing adequate professional drug administration monitoring support, (3) an infrastructure for delivery, dosing, and disposal is in place, and (4) the parenteral medication is easy to administer and preferably available in a ready-to-use injection device.

Regulators view the approval of home- and/or self-administration from a risk management perspective. In the absence of concrete guidelines, the discussions and decisions follow from risks identified or anticipated by regulators when reviewing the data package for each drug during the assessment process. Risk considerations include whether home- and self-administration adds to the overall risks identified for the biological molecule from clinical studies and administration in a controlled setting. Moreover, the nature of the identified or expected additional risks and how these risks can be mitigated needs to be assessed. Risks inherent to dosing outside of a controlled setting typically relate to the safety of the drug or drug class, the product's stability at different temperatures, and the complexity of handling and administration.

To unfold the full potential of a flexible care setting in oncology, eligible molecules have to become available in a product presentation that is easy to use and does not pose logistical barriers to drug supply, storage, and disposal. Learnings from other disease areas can be applied and it is expected that PFS, PFP, and AI technology platforms used in different indications can also be applied to low-volume subcutaneous formulations. For the higher volume formulations, the development of OBDSs is warranted to enable caregiver-assisted home- and/or patient self-administration. Connected devices and accompanying health apps with dosing reminders, adherence trackers, and patient diaries to collect patient-reported outcomes need to be assessed for their potential in supporting at-home dosing while maintaining a remote contact with the treating physician.

Overall, and especially in light of the COVID-19 pandemic, efforts in formulation and device lifecycle management that focus on simplifying drug delivery and putting patients' needs in the center will drastically accelerate, also in cancer care. Now that the value of reducing the dosing complexity for parenterally administered biotherapeutics is broadly recognized, related efforts are more and more acknowledged and supported. Formulation and device technologies that were considered "nice-to-have" until very recently have now transformed into a "must-have."

References

Alshamrani, M., AlHarbi, A., Alkhudair, N., AlNajjar, F., Khan, M., Obaid, A. B., & AlNahedh, M. (2020). Practical strategies to manage cancer patients during the COVID-19 pandemic: Saudi oncology pharmacy assembly experts recommendations. *Journal of Oncology Pharmacy Practice*, *26*(6), 1429–1440. Available from https://doi.org/10.1177/1078155220935564.

Al-Shamsi, H. O., Alhazzani, W., Alhuraiji, A., Coomes, E. A., Chemaly, R. F., Almuhanna, M., & Xie, C. (2020). A practical approach to the management of cancer patients during the novel coronavirus disease 2019 (COVID-19) pandemic: an international collaborative group. *The Oncologist*, *25*(6), e936–e945. Available from https://doi.org/10.1634/theoncologist.2020-0213.

Assouline, S., Buccheri, V., Delmer, A., Gaidano, G., McIntyre, C., Brewster, M., & Badoux, X. (2015). Pharmacokinetics and safety of subcutaneous rituximab plus fludarabine and cyclophosphamide for patients with chronic lymphocytic leukaemia. *British Journal of Clinical Pharmacology*, *80*(5), 1001–1009. Available from https://doi.org/10.1111/bcp.12662.

Barrios, V., Cosín-Sales, J., Bravo, M., Escobar, C., Gámez, J. M., Huelmos, A., & Cequier, Á. (2020). Telemedicine consultation for the clinical cardiologists in the era of COVID-19: present and future. Consensus document of the Spanish Society of Cardiology. *Revista Española de Cardiología (English Edition)*, *73*(11), 910–918. Available from https://doi.org/10.1016/j.rec.2020.06.032.

Basic, E., Kappel, M., Misra, A., Sellner, L., Ratsch, B. A., & Ostwald, D. A. (2020). Budget impact analysis of the use of oral and intravenous therapy regimens for the treatment of relapsed or refractory multiple myeloma in Germany. *The European Journal of Health Economics*, *21*(9), 1351–1361. Available from https://doi.org/10.1007/s10198-020-01219-3.

Bateman, J., & Ferner, R. (2006). Anaphylaxis: clinical features, management and avoidance. *Prescriber*, *17*(10), 12–18. Available from https://doi.org/10.1002/psb.378.

Bhatla, N., & Singhal, S. (2020). The COVID-19 pandemic and implications for gynaecologic cancer care. *Indian Journal of Gynecologic Oncology*, *18*(2), 48. Available from https://doi.org/10.1007/s40944-020-00395-7.

Biagioli, V., Albanesi, B., Belloni, S., Piredda, A., & Caruso, R. (2020). Living with cancer in the COVID-19 pandemic: an Italian survey on self-isolation at home. *European Journal of Cancer Care*, *30*(2), e13385. Available from https://doi.org/10.1111/ecc.13385.

Bittner, B., Richter, W., Hourcade-Potelleret, F., McIntyre, C., Herting, F., Zepeda, M., & Schmidt, J. (2012). Development of a subcutaneous formulation for trastuzumab – nonclinical and clinical bridging approach to the approved intravenous dosing regimen. *Arzneimittelforschung*, *62*(9), 401–409. Available from https://doi.org/10.1055/s-0032-1321831.

Bittner, B., Richter, W., & Schmidt, J. (2018). Subcutaneous administration of biotherapeutics: An overview of current challenges and opportunities. *BioDrugs*, *32*(5), 425–440. Available from https://doi.org/10.1007/s40259-018-0295-0.

Bittner, B., & Schmidt, J. (2021). In E. Chappel (Ed.), *Subcutaneous drug delivery devices – enablers of a flexible care setting* (pp. 159–179). Elsevier. Available from https://doi.org/10.1016/B978-0-12-819838-4.00021-3.

Blincyto. European Public Assessment Report. (2021). https://www.ema.europa.eu/en/documents/medication-error/blincyto-measures-avoid-medication-errors_en.pdf.

Blincyto: highlights of prescribing information. (2021). https://www.accessdata.fda.gov/drugsatfda_docs/label/2021/125557s020lbl.pdf.

Blincyto: risk evaluation and mitigation strategy (REMS). (2021). https://www.blincytorems.com/.

Blincyto: risk assessment and risk mitigation review(s). (2014). https://www.accessdata.fda.gov/drugsatfda_docs/nda/2014/125557Orig1s000RiskR.pdf.

Blincyto: summary of product characteristics. (2018). https://www.ema.europa.eu/en/documents/product-information/blincyto-epar-product-information_en.pdf.

Blincyto: summary of risk management plan. (2015). https://www.ema.europa.eu/en/documents/rmp-summary/blincyto-epar-risk-management-plan-summary_en.pdf.

Bordonaro, S., Romano, F., Lanteri, E., Cappuccio, F., Indorato, R., Butera, A., & Tralongo, P. (2014). Effect of a structured, active, home-based cancer-treatment program for the management of patients on oral chemotherapy. *Patient Preference and Adherence*, *8*, 917–923. Available from https://doi.org/10.2147/PPA.S62666.

Chan, H. W., Russell, A. M., & Smith, M. Y. (2018). What is the quality of drug safety information for patients: An analysis of REMS educational materials. *Pharmacoepidemiology and Drug Safety*, *27*(9), 969–978. Available from https://doi.org/10.1002/pds.4614.

Chaplin, S. (2020). Galcanezumab: a monoclonal antibody for migraine prophylaxis. *Prescriber*, *31*(2), 33–34. Available from https://doi.org/10.1002/psb.1826.

Ciftci, B., & Avsar, G. (2016). Assessment of three sites in terms of bruising in subcutaneous heparin administration. *International Journal of Caring Sciences*, *9*(3), 923–931. Available from http://www.internationaljournalofcaringsciences.org/docs/21_ciftsi_originial_9_3.pdf.

Cool, L., Missiaen, J., Vandijck, D., Lefebvre, T., Lycke, M., De Jonghe, P. J., & Van Eygen, K. (2019). An observational pilot study to evaluate the feasibility and quality of oncological home-hospitalization. *European Journal of Oncology Nursing*, *40*, 44–52. Available from https://doi.org/10.1016/j.ejon.2019.03.003.

Cowan, J., Bonagura, V. R., Lugar, P. L., Maglione, P. J., Patel, N. C., Vinh, D. C., & Rojavin, M. A. (2021). Safety and tolerability of manual push administration of subcutaneous IgPro20 at high infusion rates in patients with primary immunodeficiency: Findings from the manual push administration cohort of the HILO study. *Journal of Clinical Immunology*, *41*, 66–75. Available from https://doi.org/10.1007/s10875-020-00876-6.

Davies, A., Berge, C., Boehnke, A., Dadabhoy, A., Lugtenburg, P., Rule, S., & Badoux, X. (2017). Subcutaneous rituximab for the treatment of B-cell hematologic malignancies: a review of the scientific rationale and clinical development. *Advances in Therapy*, *34*, 2210–2231. Available from https://doi.org/10.1007/s12325-017-0610-z.

Davies, A., Merli, F., Mihaljevic, B., Siritanaratkul, N., Solal-Céligny, P., Barrett, M., & MacDonald, D. (2014). Pharmacokinetics and safety of subcutaneous rituximab in follicular lymphoma (SABRINA): Stage 1 analysis of a randomised phase 3 study. *The Lancet Oncology*, *15*(3), 343–352. Available from https://doi.org/10.1016/S1470-2045(14)70005-1.

Dent, S., Ammendolea, C., Christofides, A., Edwards, S., Incekol, D., Pourmirza, B., & Poirier, B. (2019). A multidisciplinary perspective on the subcutaneous administration of trastuzumab in HER2-positive breast cancer. *Current Oncology*, *26*(1), e70–e80. Available from https://doi.org/10.3747/co.26.4220.

Dowd, B. E., & Laugesen, M. J. (2020). Fee-for-service payment is not the (main) problem. *Health Services Research*, *55*(4), 491–495. Available from https://doi.org/10.1111/1475-6773.13316.

EMC – SmPC database. (2021). https://www.medicines.org.uk/emc/.

Emiliano, W., Telhada, J., & do Sameiro Carvalho, M. (2017). Home health care logistics planning: A review and framework. *Procedia Manufacturing*, *13*, 948–955. Available from https://doi.org/10.1016/j.promfg.2017.09.165.

European Public Assessment Report. (2021). https://www.ema.europa.eu/en/documents/medication-error/blincyto-measures-avoid-medication-errors_en.pdf.

Foley, R., Kaedbey, R., Song, K., Venner, C. P., White, D., Doucette, S., & Reece, D. E. (2020). Canadian perspective on managing multiple myeloma during the covid-19 pandemic: Lessons learned and future considerations. *Current Oncology*, *27*(5), 270–274. Available from https://doi.org/10.3747/co.27.7149.

Geurts-Voerman, G. E., Verhoef, L. M., Van Den Bemt, B. J. F., & Den Broeder, A. A. (2020). The pharmacological and clinical aspects behind dose loading of biological disease modifying anti-rheumatic drugs (bDMARDs) in auto-immune rheumatic diseases (AIRDs): Rationale and systematic narrative review of clinical evidence. *BMC Rheumatology*, *4*, 37. Available from https://doi.org/10.1186/s41927-020-00130-x.

Grace, E., Goldblum, O., Renda, L., Agada, N., See, K., Leonardi, C., & Menter, A. (2019). Injection site reactions in the federal adverse event reporting system (FAERS) post-marketing database vary among biologics approved to treat moderate-to-severe psoriasis. *Dermatology and Therapy*, *10*(1), 99–106. Available from https://doi.org/10.1007/s13555-019-00341-2.

Gu, Y., Zalkikar, A., Kelly, L., Daly, K., & Ward, T.E. (2020). Predicting injectable medication adherence via a smart sharps bin and machine learning. https://arxiv.org/pdf/2004.01144.pdf.

Handley, Nathan R., & Bekelman, J. E. (2019). The oncology hospital at home. *Journal of Clinical Oncology*, *37*(6), 448–452. Available from https://doi.org/10.1200/JCO.18.01167.

Hanna, K. S., Segal, E. M., Barlow, A., & Barlow, B. (2021). Clinical strategies for optimizing infusion center care through a pandemic. *Journal of Oncology Pharmacy Practice*, *27*(1), 165–179. Available from https://doi.org/10.1177/1078155220960211.

Hernandez, I., Bott, S. W., Patel, A. S., Wolf, C. G., Hospodar, A. R., Sampathkumar, S., & Shrank, W. H. (2018). Pricing of monoclonal antibody therapies: Higher if used for cancer? *American Journal of Managed Care*, *24*(2), 109–112. Available from https://pubmed.ncbi.nlm.nih.gov/29461857/.

Hoffmann, F. A., Trenova, A., Llaneza, M. A., Fischer, J., Lus, G., von Bredow, D., & Bakshi, R. (2017). Patient satisfaction with ExtaviPro™ 30G, a new auto-injector for administering interferon β-1b in multiple sclerosis: Results from a real-world, observational EXCHANGE study. *BMC Neurology*, *17*(1), 156. Available from https://doi.org/10.1186/s12883-017-0928-9.

Hofmarcher, T., Lindgren, P., Wilking, N., & Jönsson, B. (2020). The cost of cancer in Europe 2018. *European Journal of Cancer*, *129*, 41–49. Available from https://doi.org/10.1016/j.ejca.2020.01.011.

Hwang, E. S., Balch, C. M., Balch, G. C., Feldman, S. M., Golshan, M., Grobmyer, S. R., & Tanabe, K. K. (2020). Surgical oncologists and the COVID-19 pandemic: Guiding cancer patients effectively through turbulence and change. *Annals of Surgical Oncology*, *27*, 2600–2613. Available from https://doi.org/10.1245/s10434-020-08673-6.

Isabwe, G. A. C., Garcia Neuer, M., de las Vecillas Sanchez, L., Lynch, D. M., Marquis, K., & Castells, M. (2018). Hypersensitivity reactions to therapeutic monoclonal antibodies: phenotypes and endotypes. *Journal of Allergy and Clinical Immunology*, *142*(1), 159–170, e2. Available from https://doi.org/10.1016/j.jaci.2018.02.018.

Ismael, G., Hegg, R., Muehlbauer, S., Heinzmann, D., Lum, B., Kim, S. B., & Jackisch, C. (2012). Subcutaneous versus intravenous administration of (neo)adjuvant trastuzumab in patients with HER2-positive, clinical stage I-III breast cancer (HannaH study): A phase 3, open-label, multicentre, randomised trial. *The Lancet Oncology*, *13*(9), 869–878. Available from https://doi.org/10.1016/S1470-2045(12)70329-7.

Kirschbrown, W. P., Wynne, C., Kågedal, M., Wada, R., Li, H., Wang, B., & Garg, A. (2019). Development of a subcutaneous fixed-dose combination of pertuzumab and trastuzumab: Results from the phase Ib dose-finding study. *Journal of Clinical Pharmacology*, *59*(5), 702–716. Available from https://doi.org/10.1002/jcph.1362.

Laptoš, T., & Omersel, J. (2018). The importance of handling high-value biologicals: Physico-chemical instability and immunogenicity of monoclonal antibodies (review). *Experimental and Therapeutic Medicine*, *15*(4), 3161–3168. Available from https://doi.org/10.3892/etm.2018.5821.

Laughlin, A. I., Begley, M., Delaney, T., Zinck, L., Schuchter, L. M., Doyle, J., & Scott, C. A. (2020). Accelerating the delivery of cancer care at home during the Covid-19 pandemic. *NEJM Catalyst*. Available from https://doi.org/10.1056/CAT.20.0258.

Lavallee, D. C., Lee, J. R., Austin, E., Bloch, R., Lawrence, S. O., McCall, D., & Amtmann, D. (2020). mHealth and patient generated health data: Stakeholder perspectives on opportunities and barriers for transforming healthcare. *MHealth*, *6*(8). Available from https://doi.org/10.21037/mhealth.2019.09.17.

Lindsey, S., Parsons, L. B., Figg, L. R., & Rhodes, J. (2018). Evaluation of the dosing strategies of biologic agents and the theoretical impact of dose rounding. *Journal of Oncology Pharmacy Practice*, *24*(1), 47–55. Available from https://doi.org/10.1177/1078155216675518.

Makki, M., Hassali, M. A., Awaisu, A., & Hashmi, F. (2019). The prevalence of unused medications in homes. *Pharmacy (Basel)*, 7(2), 61. Available from https://doi.org/10.3390/pharmacy7020061.

Malard, F., & Mohty, M. (2020). Management of patients with multiple myeloma during the COVID-19 pandemic. *The Lancet Haematology*, 7(6), e435–e437. Available from https://doi.org/10.1016/S2352-3026(20)30124-1.

Mateos, M. V., Nahi, H., Legiec, W., Grosicki, S., Vorobyev, V., Spicka, I., & Usmani, S. Z. (2020). Subcutaneous versus intravenous daratumumab in patients with relapsed or refractory multiple myeloma (COLUMBA): A multicentre, open-label, non-inferiority, randomised, phase 3 trial. *The Lancet Haematology*, 7(5), e370–e380. Available from https://doi.org/10.1016/S2352-3026(20)30070-3.

Meuse, D. (2020). Is COVID-19 the end of fee-for-service payment? https://www.shvs.org/is-covid-19-the-end-of-fee-for-service-payment/.

Newman, M. J., & Benani, D. J. (2016). A review of blinatumomab, a novel immunotherapy. *Journal of Oncology Pharmacy Practice*, *22*(4), 639–645. Available from https://doi.org/10.1177/1078155215618770.

NICE – National Institute for Health and Care Excellence. (2017). Blinatumomab for previously treated Philadelphia-chromosome-negative acute lymphoblastic leukaemia. Technology Appraisal Guidance. https://www.nice.org.uk/guidance/ta450/resources/blinatumomab-for-previously-treated-philadelphiachromosomenegative-acute-lymphoblastic-leukaemia-pdf-82604838372037.

Novakovic, A. M., Wilkins, J. J., Dai, H., Wade, J. R., Neuteboom, B., Brar, S., & Khandelwal, A. (2019). Changing body weight-based dosing to a flat dose for avelumab in metastatic Merkel Cell and advanced urothelial carcinoma. *Clinical Pharmacology & Therapeutics*, *107*(3), 588–596. Available from https://doi.org/10.1002/cpt.1645.

Phillips, J. L., & Currow, D. C. (2010). Cancer as a chronic disease. *Collegian*, *17*(2), 47–50. Available from https://doi.org/10.1016/j.colegn.2010.04.007.

Pouls, B. P. H., Kristensen, L. E., Petersson, M., van den Bemt, B. J. F., Ballerini, L., Bruggraber, R., & Jørgensen, T. S. (2020). A pilot study examining patient preference and satisfaction for ava, a reusable electronic injection device to administer certolizumab pegol. *Expert Opinion on Drug Delivery*, *17*(5), 705–711. Available from https://doi.org/10.1080/17425247.2020.1736552.

Rombouts, M. D., Swart, E. L., Van Den Eertwegh, A. J. M., & Crul, M. (2020). Systematic review on infusion reactions to and infusion rate of monoclonal antibodies used in cancer treatment. *Anticancer Research*, *40*(3), 1201–1218. Available from https://doi.org/10.21873/anticanres.14062.

Roselló, S., Blasco, I., Fabregat, L. G., Cervantes, A., & Jordan, K.ESMO Guidelines Committee. (2017). Management of infusion reactions to systemic anticancer therapy: ESMO clinical practice guidelines. *Annals of Oncology*, *28*(Supplement 4), iv100–iv118. Available from https://doi.org/10.1093/annonc/mdx216.

Stein, A., Franklin, J. L., Chia, V. M., Arrindell, D., Kormany, W., Wright, J., & Joseph, G. (2019). Benefit-risk assessment of blinatumomab in the treatment of relapsed/refractory B-cell precursor acute lymphoblastic leukemia. *Drug Safety*, *42*(5), 587–601. Available from https://doi.org/10.1007/s40264-018-0760-1.

Summary of product characteristics – European Medicines Agency (EMA). (2021). https://www.ema.europa.eu/en/glossary/summary-product-characteristics.

Tan, A. R., Im, S. A., Mattar, A., Colomer, R., Stroyakovskii, D., Nowecki, Z., & Jackisch, C.FeDeriCa study group. (2021). Fixed-dose combination of pertuzumab and trastuzumab for subcutaneous injection plus chemotherapy in HER2-positive early breast cancer (FeDeriCa): A randomised, open-label, multicentre, non-inferiority, phase 3 study. *The Lancet Oncology*, *22*(1), 85–97. Available from https://doi.org/10.1016/S1470-2045(20)30536-2.

Tatla, D., Mountian, I., Szegvari, B., VanLunen, B., & Schiff, M. (2020). A multicenter, open-label study to evaluate the safe and effective use of a new electromechanical auto-injection device for self-injection of certolizumab pegol. *Expert Opinion on Drug Delivery*, *17*(6), 855–862. Available from https://doi.org/10.1080/17425247.2020.1747430.

Tetteh, E. K., & Morris, S. (2014). Evaluating the administration costs of biologic drugs: Development of a cost algorithm. *Health Economics Review*, *4*(1), 26. Available from https://doi.org/10.1186/s13561-014-0026-2.

Tralongo, P., Ferraù, F., Borsellino, N., Verderame, F., Caruso, M., Giuffrida, D., & Gebbia, V. (2011). Cancer patient-centered home care: A new model for health care in oncology. *Therapeutics and Clinical Risk Management*, *7*, 387–392. Available from https://doi.org/10.2147/TCRM.S22119.

Turner, P. J., Worm, M., Ansotegui, I. J., El-Gamal, Y., Rivas, M. F., Fineman, S., & Cardona, V.WAO Anaphylaxis Committee. (2019). Time to revisit the definition and clinical criteria for anaphylaxis? *World Allergy Organization Journal*, *12*(10), 100066. Available from https://doi.org/10.1016/j.waojou.2019.100066.

Usmani, S. Z., Nahi, H., Mateos, M. V., van de Donk, N. W. C. J., Chari, A., Kaufman, J. L., & San-Miguel, J. (2019). Subcutaneous delivery of daratumumab in relapsed or refractory multiple myeloma. *Blood*, *134*(8), 668–677. Available from https://doi.org/10.1182/blood.2019000667.

Uzzo, R.G., Kutikov, A., & Geynisman, D.M. (2020). Coronavirus disease 2019 (COVID-19): Cancer screening, diagnosis, treatment, and posttreatment surveillance in uninfected patients during the pandemic. UpToDate. https://www.uptodate.com/contents/coronavirus-disease-2019-covid-19-cancer-screening-diagnosis-treatment-and-posttreatment-surveillance-in-uninfected-patients-during-the-pandemic.

Vaartio-Rajalin, H., & Fagerström, L. (2019). Professional care at home: patient-centredness, interprofessionality and effectivity? A scoping review. *Health and Social Care in the Community*, *27*(4), e270–e288. Available from https://doi.org/10.1111/hsc.12731.

Vaartio-Rajalin, H., Näsman, Y., & Fagerström, L. (2020). Nurses' activities and time management during home healthcare visits. *Scandinavian Journal of Caring Sciences*, *34*(4), 1045–1053. Available from https://doi.org/10.1111/scs.12813.

Waddill, K. (2019). Proposed rule to expand value-based payment in home healthcare. https://healthpayerintelligence.com/news/proposed-rule-to-expand-value-based-payment-in-home-healthcare.

Wang, D. D., Zhang, S., Zhao, H., Men, A. Y., & Parivar, K. (2009). Fixed dosing versus body size-based dosing of monoclonal antibodies in adult clinical trials. *The Journal of Clinical Pharmacology*, *49*(9), 1012–1024. Available from https://doi.org/10.1177/0091270009337512.

Wasserman, R. L. (2017). Recombinant human hyaluronidase-facilitated subcutaneous immunoglobulin infusion in primary immunodeficiency diseases. *Immunotherapy*, *9* (12), 1035–1050. Available from https://doi.org/10.2217/imt-2017-0092.

Wynne, C., Harvey, V., Schwabe, C., Waaka, D., McIntyre, C., & Bittner, B. (2013). Comparison of subcutaneous and intravenous administration of trastuzumab: A phase I/Ib trial in healthy male volunteers and patients with HER2-positive breast cancer. *The Journal of Clinical Pharmacology*, *53*(2), 192–201. Available from https://doi.org/10.1177/0091270012436560.

CHAPTER 5

Formulation and device lifecycle management of biotherapeutics—leveraging synergies across therapeutic areas

Introduction

The previous chapters provided insights into the scientific basis for the development pathway for subcutaneous formulations for monoclonal antibodies (mAbs) (Chapter 2), described the underlying concepts for clinical bridging approaches supporting a change from manual or semiautomated subcutaneous injection to automated injection by means of a standardized device technology (Chapter 3), and provided an overview on current development activities aiming at enabling subcutaneous drug administration in a decentralized setting for biological agents in cancer care (Chapter 4).

The above product optimizations are conducted to improve patient experience with the dosing regimen and in response to the increasing pressure on healthcare costs and resources globally. Moreover, in light of the COVID-19 pandemic, the value of drug delivery improvements is increasingly recognized. Especially for biopharmaceuticals that need to be administered parenterally, reducing the dosing complexity via optimized product presentations can be an important measure to increase patient experience and access to medications globally (Okegbe et al., 2019).

Reductions in healthcare expenditure with, for example, subcutaneous versus intravenous administration of biotherapeutics have been realized in different countries. Although drug administration-related expenditures represent a comparatively small proportion of the total treatment costs, switching from intravenous to subcutaneous administration generates savings by increasing institutional capacity due to the shorter infusion duration, particularly when considering the large number of patients receiving biological treatments with multiple administrations (Rojas et al., 2020; Wallace et al., 2020).

Formulation and Device Lifecycle Management of Biotherapeutics
DOI: https://doi.org/10.1016/B978-0-12-823741-0.00005-1
© 2022 Elsevier Inc.
All rights reserved.

Consequently, to enable the timely availability of convenient and uncomplicated dosing regimens, manufacturers have to assess the feasibility of underlying product optimizations early during the lifecycle of a biological molecule. Especially, as aspects of nonclinical and clinical development, technical feasibility, as well as the applicability of formulation and device technology platforms follow the same principles independent of the disease area, related learnings from other indications should be taken into consideration to streamline development and thus ideally offer an optimally customized product presentation already with the initial launch of the molecule.

Synergies between different indications and products span from experiences on the technical feasibility of novel formulation or device technologies, the applicability of pharmacokinetic-based nonclinical and clinical bridging approaches for new administration routes, to patient and healthcare provider preferences for one over another drug delivery modality.

To apply previous learning experiences from product optimizations for molecules developed in different indications, interdisciplinary scientists, clinicians, and commercial managers need to collaborate and share available insights and best practices with their counterparts from other disease areas.

Leveraging cross-indication and cross-product synergies to accelerate formulation and device lifecycle management

The focus of this chapter is on how manufacturers can leverage cross-indication and cross-product synergies with the aim to accelerate formulation and device lifecycle management. In scope are product optimizations for biological molecules that are either in late-stage development or on the market with a different drug delivery profile or product presentation.

When initiating formulation and device lifecycle management for a biological medication, the main focus is on providing a customer-centric product presentation according to patient and healthcare provider needs, preferences, and capabilities. Manufacturers need to demonstrate that the efficacy and safety profile with the optimized product is not inferior to that of the initial presentation.

Product optimizations may be conducted to change the treatment setting, that is, the place of drug administration, based on an emerging standard-of-care. This includes new combination partners that could be dosed in a noncentralized setting as opposed to drug administration in the clinic. Aiming at parenteral home- and patient self-administration, the technical feasibility to develop, for example, subcutaneous formulations, ideally presented in ready-to-use devices, has to be evaluated.

In addition, the impact of such simplified drug delivery options on access to medication requires careful consideration. Related assessments have to include insights on affordability of the new product for patients. Moreover, the reimbursement scheme in a given legislation should support healthcare providers in their efforts to provide remote supervision in an outpatient setting Fig. 5.1.

Patient centricity

A definition of the term "patient centricity" has been laid down by Gill in his 2020 article on the promotion of self-care. Here, patient-centric approaches are defined as "involving the patient in decisions, collaborating with the patient as a partner" (Gill, 2020). In the context of choosing the most appropriate drug delivery modality, this patient empowerment

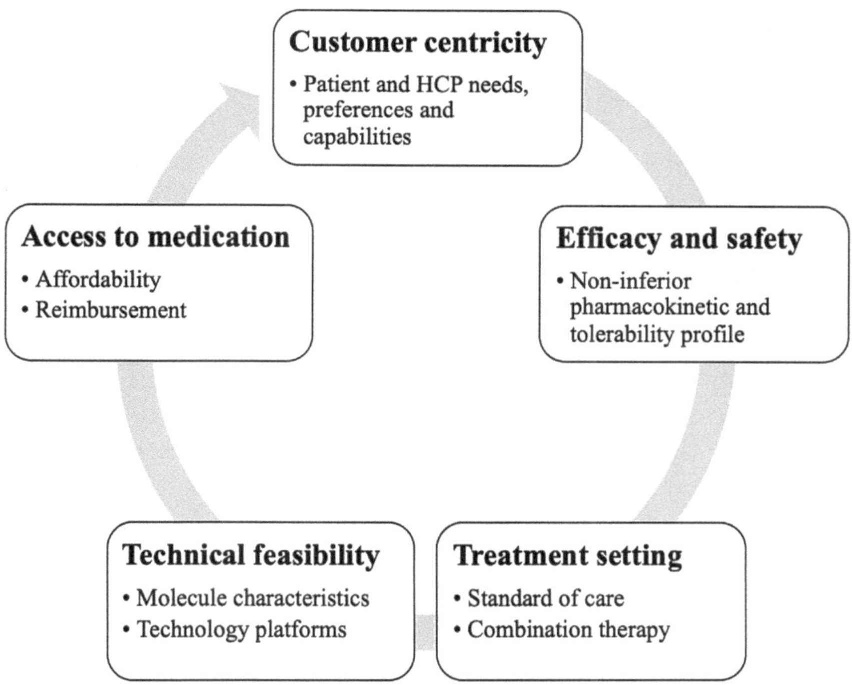

Figure 5.1 Points to consider when initiating formulation and device lifecycle management for a biological medication.

would ideally include the availability of several product presentations to account for individual desires.

Thus to envision a truly patient-centric drug administration profile, patient and healthcare professional preferences and capabilities are the key drivers for any decision on one over the other product optimization. For example, today we know that across indications, the majority of patients and healthcare providers prefer subcutaneous over intravenous administration, especially when it comes to drug administration outside of a healthcare institution (Stoner et al., 2014). Patients value subcutaneous dosing due to the perceived reduced invasiveness compared to an intravenous infusion, the shorter time required for in-clinic visits, as well as the potential for home-administration (Dychter et al., 2012; Jin et al., 2015). Physicians and nurses acknowledge advantages, such as improved healthcare resource utilization, increased flexibility in appointment scheduling, and reduced capacity bottlenecks and nursing overtime (Jackisch et al., 2015). As these points are generally independent of the indication, manufacturers can leverage available insights from primary and secondary market research on patient, physician, and nurse preferences as well as from studies on the respective health economic benefit with different routes of administration.

Despite the majority of patients and caregivers leaning towards subcutaneous as opposed to intravenous dosing, some still favor intravenous infusions and the option to be dosed in the clinic or infusion center. This can be the case for needle-phobic patients (Hamilton, 1995), patients who value personal exchange with other patients or with a healthcare professional during their visit in a healthcare institution (Browall et al., 2013), or patients who receive a treatment in combination with other intravenously dosed medications. In the latter case, delivering the two treatments intravenously via the same intravenous port can be a convenience advantage (Bittner et al., 2018). Also, special patient populations (i.e., motor or mentally impaired, children, elderly) may require individualized injection device types and the support of a healthcare provider. The related specifications for convenient subcutaneous product presentations therefore need to be assessed by the manufacturer on a case-by-case basis.

Efficacy and safety

A crucial element associated with formulation and device lifecycle management is to demonstrate that a product optimization does not negatively impact the established efficacy and safety profile of a biological drug. In the following, related synergies and differences between products will be illustrated for the example of changing from intravenous to subcutaneous administration of an mAb. For details on the underlying scientific concepts and development pathway(s), refer to Chapter 2.

The potential impact of a change from an intravenous to a subcutaneous dosing regimen on clinical outcomes and tolerability, including immunogenicity, is best visualized by comparing the pharmacokinetic profiles of an mAb following administration via either route. With the aim of achieving the same overall exposure [expressed as area under the serum concentration–time curve (AUC)], injecting a mAb intravenously and thus directly into the bloodstream results in immediate maximum serum concentrations (C_{max}). At the same AUC, the pharmacokinetic profile of subcutaneously injected mAbs is typically characterized by a slow absorption from the subcutaneous tissue with C_{max} levels below those achieved following intravenous dosing. To compensate for the higher C_{max} with intravenous administration, at the same overall exposure, the minimum or trough serum concentration (C_{trough}) with subcutaneous dosing generally exceeds that of the intravenous dose. This hypothesis has been shown to be applicable for different mAbs and builds the basis for estimating possible changes in the efficacy and safety of administration route changes independent of the therapeutic area (Bittner et al., 2018).

Notably, there are situations in which a biological can only be developed with an intravenous dosing regimen. This may be the case, for example, for antibody–drug conjugates. Here, a potent cytotoxic component is conjugated to a mAb and can cause off-target toxicities mediated by immune cells in the skin (Lucas et al., 2018). In addition, in the event that an immediate onset of action is needed, intravenous delivery would be warranted (Taussky et al., 2011).

Efficacy

For mAbs in oncology for which an intravenous dosing regimen is already available, the subcutaneous doses are selected to achieve C_{trough} and/or AUC levels noninferior to those with intravenous dosing (refer to Chapter 2). This approach is expected to result in a comparable saturation of target receptors and ultimately noninferior efficacy (Mager & Jusko, 2001). Previous comparisons of subcutaneous trastuzumab, rituximab, or daratumumab with their respective intravenous dosing regimens (Davies et al., 2014; Ismael et al., 2012; Mateos et al., 2020) revealed that C_{max} did not correlate with clinical efficacy. Therefore despite the lower C_{max} following subcutaneous administration, the risk of underexposing patients with the novel subcutaneous dosing regimen was considered low. The same dose finding concept was applied to the coformulation of trastuzumab and pertuzumab, the first fixed-dose combination of two mAbs (Kirschbrown et al., 2018).

Clinical development programs for bridging from intravenous to subcutaneous dosing for the above biologics did consist of a phase 1b dose finding study to select the subcutaneous dose expected to result in noninferior C_{trough} and AUC between the two regimens and a comparatively small phase 3 study to demonstrate noninferiority in pharmacokinetic endpoints (C_{trough}, AUC) supplemented with efficacy data (i.e., response rates) (Shpilberg & Jackisch, 2013).

This clinical development package was complemented by data from nonclinical models (Bittner et al., 2012, 2014). The nonclinical program for trastuzumab, for example, included data from a mouse xenograft model in BALB/c nu/nu mice transplanted with the nonsmall-cell lung cancer cell line Calu-3, which has high expression of HER2 (Scheuer et al., 2009). Data demonstrated similar pharmacodynamic activity of trastuzumab at comparable minimum serum concentration after intravenous and subcutaneous administration, irrespective of the administration route. For rituximab, the pharmacodynamic effects of subcutaneous rituximab were compared with those of intravenous rituximab in a xenograft model employing the Z138 human mantle cell lymphoma cell line (Tucker et al., 2006) in female SCID beige mice. The results from the xenograft study demonstrated that when similar trough concentrations of rituximab are reached by subcutaneous and intravenous administration, similar efficacy is achieved (Mao et al., 2013). Collectively, the results from the different xenograft models for trastuzumab and rituximab provided supporting evidence that, despite differences in C_{max}, subcutaneous administration resulted in similar pharmacodynamics and efficacy compared with intravenous rituximab.

Manufactures would need to deviate from this C_{trough}- or AUC-based bridging approach in case of a different pharmacokinetic–pharmacodynamic correlation for a given mAb. For instance, in the event that C_{max} had been shown to contribute to efficacy, differently designed and larger clinical programs would need to be conducted.

Safety

Comparing the safety and tolerability of subcutaneous versus intravenous dosing of mAbs, in general, the focus is on immunogenicity, infusion-related reactions, and injection site reactions. It is suggested in the literature that immunogenicity, that is, the formation of antidrug antibodies against a foreign biological molecule may be increased following subcutaneous injection compared to intravenous infusion. This hypothesis is based on the observation that biotherapeutics with the molecular weights of >20 kDa exhibit limited transport into the blood capillaries and cross into the circulation system predominantly via the lymphatics (McLennan et al., 2005; Turner & Balu-Iyer, 2018). To explore the impact of antidrug antibodies formed following either administration route on clinical outcomes, a possible correlation with the pharmacokinetic profile, efficacy, and tolerability of a mAb needs to be established. Such assessments are required for each individual molecule as well as for different stages of a disease. For example, patients with an autoimmune disease may show higher immunogenicity due to the inflammatory status of their immune system (Jawa et al., 2020).

In the pivotal phase 3 HannaH study for trastuzumab that compared pharmacokinetics and efficacy of trastuzumab following subcutaneous versus intravenous administration, pharmacokinetics, and pathological complete response rates within each treatment arm were similar regardless of trastuzumab antidrug antibody status or treatment cycle. In addition, the incidence of administration-related reactions was not statistically different independent of the antidrug antibody status (Hegg et al., 2012). Similar findings were made in the corresponding phase 3 SABRINA trial supporting development of subcutaneous rituximab. In patients tested positive for rituximab antidrug antibodies, no impact on B-cell depletion, efficacy, pharmacokinetics, or safety was found (Davies et al., 2017). The phase 1b PAVO study evaluated subcutaneously administered daratumumab at doses of 1200 or 1800 mg in patients with relapsed or refractory multiple myeloma. Similar to studies of intravenous daratumumab, there was a low incidence of antidaratumumab antibodies (1.3%), indicating a low risk of daratumumab immunogenicity with subcutaneous infusion (Usmani et al., 2019). As the antidrug antibody profile is different for each individual mAb, the respective development programs for different administration routes have to comprise a comprehensive data set each to exclude and interpret differences in immunogenicity and its clinical relevance.

Infusion-related reactions, characterized by flushing, rash, fever, rigors, chills, dyspnea, mild, and/or severe hypotension with(out) bronchospasms, cardiac dysfunction, and/or anaphylaxis are commonly observed following parenteral administration of many mAbs and typically occur during the first cycles of dosing (Gülsen et al., 2020; Rombouts et al., 2020). Reactions vary in quantity and intensity on a molecule-specific basis (Cáceres et al., 2019). There is some evidence that these events can be C_{max} related, so that to reduce infusion-related reactions, the infusion rate is typically slowed down (Rombouts et al., 2020).

Based on this observation, in theory, changing from intravenous to subcutaneous dosing, with lower C_{max} values at comparable overall exposure may be advantageous with respect to the incidence and severity of infusion-related reactions. Biotherapeutics, such as alemtuzumab (Campath-H1), daratumumab, or rituximab, are examples of mAbs that can be administered intravenously and subcutaneously with a comparatively high incidence of infusion-related reactions during the first intravenous infusion (Cáceres et al., 2019). While for rituximab the subcutaneous dosing regimen still includes the first dose to be infused intravenously, for alemtuzumab and daratumumab, the subcutaneous dose is given from the first cycle onwards.

For both alemtuzumab and daratumumab, there is initial evidence that this C_{max} hypothesis may hold true. Namely, a retrospective comparison of intravenous versus subcutaneous administration of alemtuzumab for graft-versus-host disease prophylaxis in unrelated donor transplant recipients revealed a significant reduction in grade ≥ 2 infusion-related reactions after subcutaneous injections. Grade ≥ 2 infusion-related reactions occurred in 11 (8%) patients in the subcutaneous and in 25 (25%) patients in the intravenous cohorts ($P = .001$). Twelve injections (9%) and 26 infusions (26%) experienced infusion-related reactions of any grade ($P = .001$) in the subcutaneous and intravenous cohorts, respectively (Patel et al., 2016).

In the phase 3 study COLUMBA, subcutaneous daratumumab resulted in significantly less infusion-related reactions compared to intravenous daratumumab, that is, 13% versus 34% (odds ratio 0.28, 95% confidence interval 0.18–0.44, $P < .0001$), respectively. The majority of infusion-related reactions occurred with the first dose and was predominantly grade 1 and 2. Grade 3 infusion-related reactions occurred in 2% in the subcutaneous compared to 5% in the intravenous cohort (Mateos et al., 2020). As per daratumumab's Summary of Product Characteristics

(The Electronic Medicines Compendium, 2021), including terms determined by investigators as related to infusion/injection of daratumumab, infusion-related reactions occur very commonly (≥1/10) following both administration routes. Infusion-related reactions of any grade were observed in 11% and 39% of patients and of grades 3–4 in 1% and 11% (no grade 4) of patients with the subcutaneous and intravenous formulations, respectively. These frequencies are based on daratumumab subcutaneous ($n = 490$) and intravenous ($n = 2324$) studies (The Electronic Medicines Compendium, 2021).

Overall, while these assessments indicate a reduction in the incidence and severity of infusion-related reactions for alemtuzumab and daratumumab following subcutaneous administration, findings cannot be generalized for other biological medications. Associated data assessments require separate analyses for each particular biological molecule.

Injection site reactions, such as local pain, itching, erythema, or induration, are frequently observed transient and reversible adverse events and predominantly occur following subcutaneous administration (Thomaidou & Ramot, 2019). As a consequence, subcutaneous injections are often given more frequently in smaller individual doses than intravenous infusions (i.e., tocilizumab is administered weekly or every 2 weeks subcutaneously versus monthly intravenously) (The Electronic Medicines Compendium, 2021). Especially, higher volume subcutaneous injections can raise concerns about local tolerability (Shapiro, 2010; Wright & Jones, 2017). To overcome the need to dose subcutaneous formulations in shorter intervals than their intravenous equivalent to avoid induration at the injection site, high-volume subcutaneous treatment can be facilitated by preinjection or coformulation with the dispersion enhancer hyaluronidase (Bittner et al., 2018). However, independent of the use of a permeation enhancer, the nature and severity of injection site reactions varies depending on the individual molecule (Scherer et al., 2010) and in some cases also depending on the composition of the dosing vehicle (Singh et al., 2018). The large available database on injection site reactions across biotherapeutics and indications provides a strong basis for scientists to assess what findings can generally be attributed to subcutaneous dosing and which are more related to the specific molecule and its mode of action.

The evolving knowledge base of the impact of the administration route on the overall tolerability and safety profile of biological medications should be used by safety scientists in evaluating what findings are molecule-specific and what findings may be attributed to the administration route-dependent absorption pathway into the systemic circulation.

Treatment setting

An important consideration when assessing the advantages of one over another administration route is the current and emerging treatment setting, that is, the way molecules are administered in a given indication. A sound understanding of the disease area-specific drug administration landscape is warranted for manufacturers to be able to address emerging needs of patients, providers, and the overall healthcare system.

Hypothetically, two major treatment settings that influence the selection of the most appropriate administration route and product presentation could be distinguished (Fig. 5.2). In the first assumed setting, biotherapeutics are on the market for monotherapy with both intravenous and subcutaneous dosing alternatives. Subcutaneous administration is predominantly with low injection volumes facilitated by means of autoinjector (AI) or prefilled pen (PFP) devices. Treatments have been shown to be safe and well tolerated and thus patients and providers have several administration choices and are used to self-administration. An example for such a treatment setting is rheumatoid arthritis with biological treatments currently comprising tumor necrosis factor-α inhibitors, a B-cell-depleting therapy, a selective costimulating modulator, and interleukin-6 or -1 receptor antagonists (Bittner et al., 2018).

In the second assumptive treatment setting, at the time of approval of a novel biological medication, intravenous infusions are the standard-of-care, but the number of subcutaneous formulations in development is steadily increasing. Dosing biotherapeutics in combination therapy is broadly evaluated in clinical trials and subcutaneous administration is seen as a way to reduce dosing complexity. Subcutaneous home- or self-administration, however, is not yet implemented widely. Presently, the drug delivery framework for mAbs in oncology represents an archetype for this treatment setting (Bittner et al., 2018).

On the assumption that subcutaneous administration is already an established treatment modality and preferred by the majority of patients and healthcare providers within a disease area, timely availability of a subcutaneous product presentation is key in order not to diminish the convenience benefit of established treatments. Thus unless a new medication provides a clear efficacy or safety benefit over the current standard-of-care, providing solely an intravenous formulation may not be readily accepted by patients and healthcare providers, which could in turn limit access to the medication.

Novel first- or best-in-class biotherapeutics should be developed with the administration route that enables fastest availability to patients. Unless established formulation or device platforms can be leveraged, manufacturers would typically regard product optimizations as a lifecycle management activity. This is to avoid any possible delay in initial molecule launch due to formulation or device development.

In areas with combination therapy, dosing regimens should minimize the number of injections and reduce dosing complexity overall. In case a combination partner is administered orally or subcutaneously, the typically more preferred administration routes (Huynh et al., 2014), only having an intravenous dosing option would negatively impact the convenience advantage of the combination partner. This is because intravenous dosing mostly mandates the aid of a healthcare provider and is predominantly conducted in a healthcare institution (Tjalma et al., 2017). Here, the availability of both intravenous and subcutaneous dosing alternatives has to be considered to account for the dosing schedule of the combination partner(s). The development of fixed-dose combinations with two or more molecules coformulated in the same dosing vehicle can further simplify parenteral drug administration of two molecules.

Independent of the disease area, manufacturers can leverage learnings from indications with similar drug delivery standards while considering the current and emerging treatment and technology landscape in the target disease. This approach lowers the need to generate additional insights via primary market research or patient and healthcare provider advisory boards. In the event a molecule is intended to be developed with an administration route that is unprecedented for the patient population, targeted research and input from therapeutic area experts on possible challenges and on how these could be overcome is mandated. Whether and when a product optimization can be realized is often conditional to molecule characteristics and the availability of suitable formulation or device platform technologies.

Technical feasibility of enabling technologies and device platforms

Patient and healthcare provider preferences are the most important drivers for determining the appropriate target product profile of a biological drug.

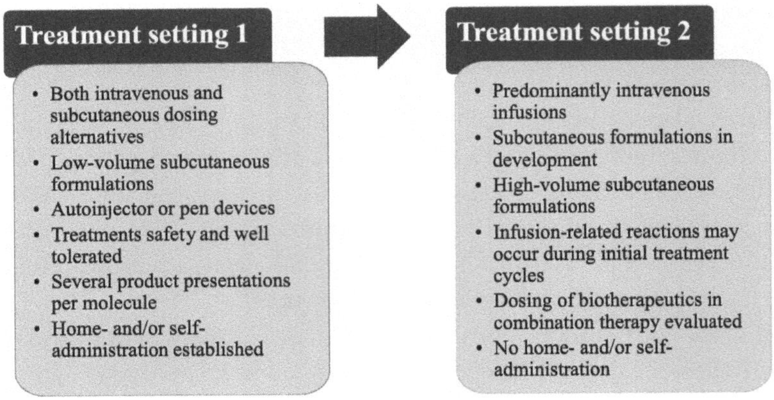

Figure 5.2 Hypothetical treatment settings that inform the selection of the most appropriate administration route and product presentation.

While today, for the development of low-dose subcutaneous ready-to-use product presentations established formulation and device technologies are available (Bittner & Schmidt, 2021), realizing a patient-centric drug administration approach for high-dose subcutaneous biologics can be challenging. Only more recently, convenient subcutaneous administration of high-dose formulations has become feasible with the development of high-concentration solutions (Tomar et al., 2016), the coformulation with the dispersion enhancer hyaluronidase to facilitate the injection of larger volumes (Frost, 2007), and the development of patch devices to facilitate a flexible care setting (Nema & Ludwig, 2019).

Table 5.1 provides an overview on prototypical subcutaneous product presentations, administration times, and treatment settings in relation to the dosing volume. To simplify the conceptual overview, the device types are classified according to the injection volume that can be delivered with the respective product presentation. Specifications should be considered as an approximate only. Notably, wearable on-body delivery systems (OBDSs) for volumes greater than 4 mL are not yet approved in combination with a biological molecule, but currently in development (Bittner & Schmidt, 2021).

Subcutaneous formulations with dosing volumes lower than approximately 2 mL are presented in a vial and injected manually with a handheld syringe (HHS) or as a ready-to-use configuration in a PFS, AI, or PFP device (Vermeire et al., 2018). In this volume range, manual injection with PFS commonly takes less than 1 min, depending on the needle

Table 5.1 Approximate administration time and treatment setting for prototypical subcutaneous product presentations for biotherapeutics aiming at fast but tolerable injection duration.

Product presentation	Range of dosing volumes (approximates only)		
	<2 mL	>2–10 mL	10–50 mL
Prefilled syringe			
Injection time	<1 min	NA	NA
Predominant treatment setting	Patient self-administration and lay caregiver/healthcare provider assisted		
Autoinjector/pen device			
Injection time	<15 s	NA	NA
Predominant treatment setting	Patient self-administration and lay caregiver/healthcare provider assisted		
Vial (manual injection with handheld syringe)			
Injection time	<1 min	1–5 min[a]	5–25 min[a]
Predominant treatment setting	Patient self-administration and lay caregiver/healthcare provider assisted	Healthcare provider assisted	Healthcare provider assisted
Wearable on-body delivery system			
Injection time	NA	1–5 min[a,b]	5–25 min[a,b]
Predominant treatment setting		Patient self-administration and lay caregiver/healthcare provider assisted	Patient self-administration and lay caregiver/healthcare provider assisted

Note: Specifications should be considered as an estimate only.
[a]Assumes an injection rate of 2 mL/min; higher injection rates may be feasible depending on the tolerability profile of the molecule.
[b]Not yet approved in combination with a biological molecule.

gage (Krayukhina et al., 2020) and the individual's preferred injection rate. A PFS facilitates patient self-administration, while manual injection with the vial presentation is predominantly conducted by a healthcare provider. Automated AI or PFP devices for self-administration mostly deliver the product within less than 15 s and are broadly favored by patients across indications (Kessler et al., 2019).

Today dosing volumes of greater than 2 mL required for mAbs, such as trastuzumab or daratumumab, are typically available in a vial requiring health-care professional-assisted administration with a HHS or infusion set (The Electronic Medicines Compendium, 2021). Injection rates vary between 2 and 5 mL/min, depending on the molecule's safety and tolerability profile, as well as on individual healthcare provider and patient preferences (Bittner et al., 2018). To date, only one high volume OBDS (West SmartDose Gen-I; 3.5 mL) has been approved for subcutaneous delivery of a biological molecule, the lipid-lowering agent evolocumab (Markarian, 2019).

A variety of subcutaneous products offer more than one device type (Bittner & Schmidt, 2021). Commonly, at initial molecule launch, products are available with only one administration route and device type. Additional dosing regimens and device options are sequentially developed as a lifecycle management activity accounting for the availability of alternative platform technologies. Fig. 5.3 illustrates three possible launch scenarios with respect to the sequence of subcutaneous administration devices.

In scenario 1, the molecule is initially introduced with an intravenous dosing regimen only and therefore requires administration by a healthcare provider within a healthcare institution. This would be the case in instances where development of a subcutaneous dosing alternative would delay initial molecule launch and thus is developed as a lifecycle management (Shpilberg & Jackisch, 2013). In the absence of a versatile device platform technology, to enable timely availability of the subcutaneous dosing alternative, the subcutaneous formulation would in a first step be marketed in a customary PFS or vial, followed by the introduction of an AI or PFP device like for tocilizumab (The Electronic Medicines Compendium, 2021) or as exemplified for evolocumab in a high-volume wearable OBDS (The Electronic Medicines Compendium, 2021). While PFSs can be applied at home by the patients themselves or with the support of the lay caregiver (Bel et al., 2020), higher volume subcutaneous dosing from the vial using a HHS is typically conducted by a healthcare professional in a controlled setting (Tetteh & Morris, 2014).

Scenario 2 illustrates a situation in which a biological agent is developed with a subcutaneous formulation from the beginning of its lifecycle. This could be imaginable for a drug in an indication with subcutaneous dosing representing an established treatment modality. To avoid a delay in approval

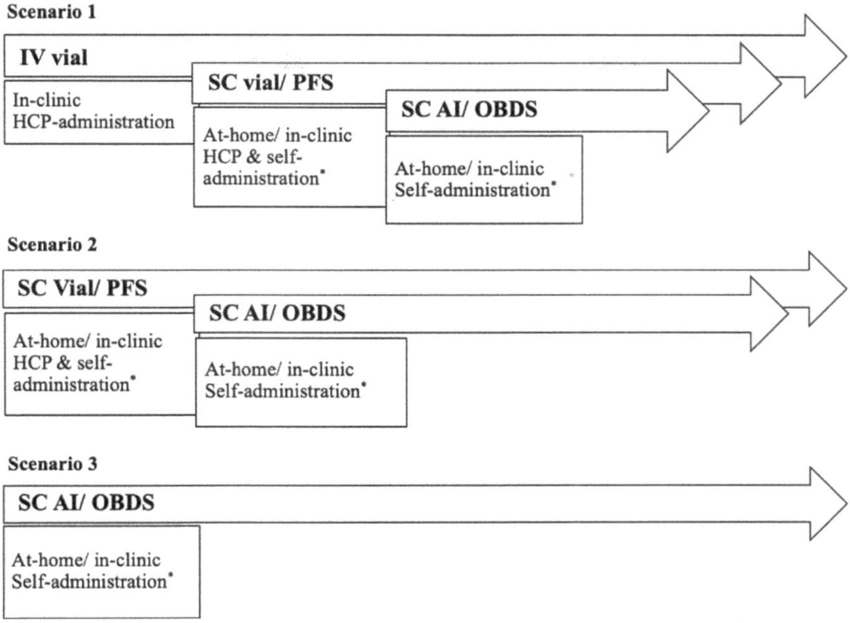

Figure 5.3 Possible launch scenarios—administration route and device type. *Patient or nonprofessional caregiver. *AI*, autoinjector; *HCP*, healthcare provider; *IV*, intravenous; *OBDS*, on-body delivery system; *PFS*, prefilled syringe; *SC*, subcutaneous.

of the medication due to the development timelines for an AI, PFP, or wearable device technology platform, the product would initially be presented in a PFS or vial followed by the automated device as a step-wise approach. Thus in this instance, patient self-administration will be possible already at initial launch for the low volume PFS. For the higher volume, self-administration would only be feasible with readiness of the wearable device.

In scenario 3, the subcutaneous formulation is directly launched with a device that enables home- and patient self-administration. At this time, scenario 3 is largely feasible for low-dose subcutaneous formulations where AI and PFP device technologies are established standard platforms. It is expected that with the evolving availability of high-volume device platforms, molecules may also be presented directly with a wearable OBDS at the time of first molecule approval.

Access to medication

Per the World Health Organization's standard definition, access is defined as "having medicines continuously available and affordable at public or private health facilities or medicine outlets that are within one hour's walk of the population" (Ekeigwe, 2019). This means that the societal benefit of a best-in-class or first-in-class medication can be limited in case the molecule is not readily accessible for a larger part of the target patient population.

Especially considering patients who live in rural regions in a distance from a healthcare institution, formulation or device technologies that allow subcutaneous administration outside of a controlled setting have the potential to shift care closer to the patient, that is, to a community center or to the patients' home. To facilitate adherence to the treatment regimen without the direct supervision of a healthcare provider, connected devices and associated health apps can be designed to enable an ongoing dialog between patients and the treating physician in a remote setting (Bittner et al., 2019; Sangave et al., 2019). To further ensure affordability of treatments, such novel formulation or device technologies should ideally be introduced without additional costs for the patient compared with already established dosing regimens. Depending on the individual patient's routines and personal reasons for nonadherence (Calvo-Alén et al., 2017), digital dosing reminders should ideally be cocreated with patients representing the range of target users (Newman-Casey et al., 2018). The insights that would support development of connected features and related health apps might not yet be available at the time of introduction of the drug delivery device. Manufacturers may therefore consider developing a number of nonpersonalized and indication-independent basic adherence tracker and dosing reminder tools. These could be applied as a solution at initial device launch and subsequently be complemented with additional features addressing specific needs of the respective patient population as a lifecycle management.

To unfold the full potential of home administration of biotherapeutics, the inherent logistical challenges and workload for healthcare providers need to be acknowledged and addressed. Consequently, a key prerequisite for the acceptability of home administration is that healthcare providers are reimbursed for their work related to remote patient care. Without incentives that support a shift of treatments to the patient's home, full realization of the value of subcutaneous administration will be challenging. Although reimbursement models vary depending on the country (EXPH, 2019), manufacturers can leverage learnings from other products on the same regional market.

Summary and conclusion

Development and commercialization of product optimizations for biotherapeutics is largely molecule-independent. Manufacturers can leverage multidisciplinary learnings from biological drugs in other disease areas that apply the same formulation and device technologies. Insights that can be translated across disease areas are manifold. A close cooperation between representatives from various scientific and development disciplines is therefore mandated to leverage learnings, but also to clearly understand differences inherent to the particular indication, the patient population, and molecule characteristics. Fig. 5.4 summarizes aspects that are generally interchangeable between different molecules and indications, as well as parameters that need to be re-assessed on a case-by-case basis depending on the physicochemical properties and mode of action of the molecule and the specific patient population.

Synergies include patients and healthcare provider preferences for one over another drug delivery methodology or administration route. Also, the health economic benefit of product optimizations, that is, the influence of alternative healthcare technologies on treatment efficiency and cost-effectiveness can to a great extent be generalized across indications. For example, technologies that facilitate parenteral dosing outside of the clinic would as a general rule reduce drug administration related institutional costs and

Molecule-/ indication-independent considerations

- Patient & provider preferences and health economic benefit
- Pharmacokinetic-based nonclinical & clinical development programs
- Technology-dependent safety & tolerability profiles
- Approaches & technologies to complement combination therapy
- Platform technologies (i.e., devices, formulation ingredients)
- Country-dependent reimbursement models on access to medication

Molecule-/ indication-dependent considerations

- Special patient populations (i.e., motor impairment, pediatric)
- Molecule-specific tolerability profile
- Current and emerging treatment & technology landscape in target indication
- Molecule-characteristics that impact technical feasibility
- Development timelines of technology platform

Figure 5.4 Molecule and indication-dependent and -independent considerations when leveraging insights from other products that apply the same formulation or device technology.

resources. Moreover, the impact of country-dependent reimbursement models that define the value of a reduced dosing complexity is to some extent transferable from one indication to another.

Alterations in the pharmacokinetic, safety, and tolerability profile of a biological molecule due to a change in the drug delivery procedure will be comparable across biotherapeuticals. The underlying knowledge can to some extent be translated to other molecules in different indications. This enables scientists in designing nonclinical and clinical development pathways for novel formulation and device technologies based on general principles while taking into account molecule-specific requirements.

Molecule-dependent aspects that guide the choice of the most appropriate administration route and device technology include special patient populations (i.e., children, elderly, motor impaired), molecule characteristics that impact the technical feasibility of a novel formulation or device technology, the molecule-specific tolerability profile, and the current and emerging treatment and technology landscape in the target indication. Related questions have to be addressed on a case-by-case basis and continuously add to the evolving knowledge base.

References

Bel, E., Bernstein., Bjermer., Follows, R., Bentley, J., Pouliquen., & Bradford, E. (2020). Usability of mepolizumab single-use prefilled syringe for patient self-administration. *Journal of Asthma*, *57*(7), 755–764. Available from https://doi.org/10.1080/02770903.2019.1604745.

Bittner, B., Chiesi, C. S., Kharawala, S., Kaur, G., & Schmidt, J. (2019). Connected drug delivery devices to complement drug treatments: Potential to facilitate disease management in home setting. *Medical Devices: Evidence and Research*, *12*, 101–127. Available from https://doi.org/10.2147/MDER.S198943.

Bittner, B., Richter, W., Hourcade-Potelleret, F., Herting, F., & Schmidt, J. (2014). Non-clinical pharmacokinetic/pharmacodynamic and early clinical studies supporting development of a novel subcutaneous formulation for the monoclonal antibody rituximab. *Drug Research*, *64*(11), 569–575. Available from https://doi.org/10.1055/s-0033-1363993.

Bittner, B., Richter, W., Hourcade-Potelleret, F., McIntyre, C., Herting, F., Zepeda, M., & Schmidt, J. (2012). Development of a subcutaneous formulation for trastuzumab—Nonclinical and clinical bridging approach to the Approved Intravenous Dosing Regimen. *Drug Research*, *62*(9), 401–409. Available from https://doi.org/10.1055/s-0032-1321831.

Bittner, B., Richter, W., & Schmidt, J. (2018). Subcutaneous administration of biotherapeutics: An overview of current challenges and opportunities. *BioDrugs*, *32*, 425–440. Available from https://doi.org/10.1007/s40259-018-0295-0.

Bittner, B., & Schmidt, J. (2021). Subcutaneous drug delivery devices—Enablers of a flexible care setting. *In Drug Delivery Devices and Therapeutic Systems*, 159–179. Available from https://doi.org/10.1016/b978-0-12-819838-4.00021-3, Elsevier.

Browall, M., Koinberg, I., Falk, H., & Wijk, H. (2013). Patients' experience of important factors in the healthcare environment in oncology care. *International Journal of Qualitative Studies on Health and Well-Being*, *8*(1). Available from https://doi.org/10.3402/qhw.v8i0.20870.

Calvo-Alén, J., Monteagudo, I., Salvador, G., Vázquez-Rodríguez, T. R., Tovar-Beltrán, J. V., Vela, P., Maceiras, F., Bustabad, S., Román-Ivorra, J. A., Díaz-Miguel, C., Rosas, J., Raya, E., Carmona, L., Cea-Calvo, L., Arteaga, M. J., Fernández, S., & Marras, C. (2017). Non-adherence to subcutaneous biological medication in patients with rheumatoid arthritis: A multicentre, non-interventional study. *Clinical and Experimental Rheumatology*, *35*(3), 423–430. Available from https://pubmed.ncbi.nlm.nih.gov/28032846/.

Cáceres, M. C., Guerrero-Martín, J., Pérez-Civantos, D., Palomo-López, P., Delgado-Mingorance, J. I., & Durán-Gómez, N. (2019). The importance of early identification of infusion-related reactions to monoclonal antibodies. *Therapeutics and Clinical Risk Management*, *15*, 965–977. Available from https://doi.org/10.2147/TCRM.S204909.

Davies, A., Merli, F., Mihaljevic, B., Siritanaratkul, N., Solal-Céligny, P., Barrett, M., Berge, C., Bittner, B., Boehnke, A., McIntyre, C., & MacDonald, D. (2014). Pharmacokinetics and safety of subcutaneous rituximab in follicular lymphoma (SABRINA): Stage 1 analysis of a randomised phase 3 study. *The Lancet Oncology*, *15*(3), 343–352. Available from https://doi.org/10.1016/S1470-2045(14)70005-1.

Davies, A., Merli, F., Mihaljević, B., Mercadal, S., Siritanaratkul, N., Solal-Céligny, P., Boehnke, A., Berge, C., Genevray, M., Zharkov, A., Dixon, M., Brewster, M., Barrett, M., & MacDonald, D. (2017). Efficacy and safety of subcutaneous rituximab vs intravenous rituximab for first-line treatment of follicular lymphoma (SABRINA): A randomised, open-label, phase 3 trial. *The Lancet Haematology*, *4*(6), e272–e282. Available from https://doi.org/10.1016/S2352-3026(17)30078-9.

Dychter, S. S., Gold, D. A., & Haller, M. F. (2012). Subcutaneous drug delivery: A route to increased safety, patient satisfaction, and reduced costs. *Journal of Infusion Nursing*, *35*(3), 154–160. Available from https://doi.org/10.1097/nan.0b013e31824d2271.

Ekeigwe, A. (2019). Drug manufacturing and access to medicines: The West African story. A literature review of challenges and proposed remediation. *AAPS Open*. Available from https://doi.org/10.1186/s41120-019-0032-x.

EXPH. (2019). Defining value in "value-based healthcare" report of the expert panel on effective ways of investing in health (EXPH). https://ec.europa.eu/health/sites/health/files/expert_panel/docs/024_defining-value-vbhc_en.pdf.

Frost, G. I. (2007). Recombinant human hyaluronidase (rHuPH20): An enabling platform for subcutaneous drug and fluid administration. *Expert Opinion on Drug Delivery*, *4*(4), 427–440. Available from https://doi.org/10.1517/17425247.4.4.427.

Gill, M. (2020). Patient centricity: Alienating patients from their own decisions is subjugation. *International Journal of Care Coordination*, *23*(4), 133–136. Available from https://doi.org/10.1177/2053434520978328.

Gülsen, A., Wedi, B., & Jappe, U. (2020). Hypersensitivity reactions to biologics (part I): Allergy as an important differential diagnosis in complex immune-derived adverse events. *Allergo Journal International*, *29*(4), 97–125. Available from https://doi.org/10.1007/s40629-020-00126-6.

Hamilton, J. G. (1995). Needle phobia: A neglected diagnosis. *Journal of Family Practice*, *41*(2), 169–175. Available from https://cdn.mdedge.com/files/s3fs-public/jfp-archived-issues/1995-volume_40-41/JFP_1995-08_v41_i2_needle-phobia-a-neglected-diagnosis.pdf.

Hegg, R., Pienkowski, T., Chen, S., Staroslawska, E., Falcon, S., Kovalenko, N., Al-Sakaff, N., Heinzmann, D., Kolaitis, G., & Ismael, G. (2012). 273P—Immunogenicity of trastuzumab intravenous and subcutaneous formulations in the Phase III HannaH study. *Annals of Oncology, 23*(Suppl. 9). Available from https://doi.org/10.1016/S0923-7534(20)32835-0, ix103.

Huynh, T., Ostergaard, A., Egsmose, C., & Madsen, O. (2014). Preferences of patients and health professionals for route and frequency of administration of biologic agents in the treatment of rheumatoid arthritis. *Patient Preference and Adherence, 8*, 93–99. Available from https://doi.org/10.2147/PPA.S55156.

Ismael, G., Hegg, R., Muehlbauer, S., Heinzmann, D., Lum, B., Kim, S. B., Pienkowski, T., Lichinitser, M., Semiglazov, V., Melichar, B., & Jackisch, C. (2012). Subcutaneous vs intravenous administration of (neo)adjuvant trastuzumab in patients with HER2-positive, clinical stage I-III breast cancer (HannaH study): A phase 3, open-label, multicentre, randomised trial. *The Lancet Oncology, 13*(9), 869–878. Available from https://doi.org/10.1016/S1470-2045(12)70329-7.

Jackisch, C., Müller, V., Dall, P., Neumeister, R., Park-Simon, T. W., Ruf-Dördelmann, A., Seiler, S., Tesch, H., & Ataseven, B. (2015). Subcutaneous trastuzumab for HER2-positive breast cancer—Evidence and practical experience in 7 German centers. *Geburtshilfe Frauenheilkunde, 75*(6), 566–573. Available from https://doi.org/10.1055/s-0035-1546172.

Jawa, V., Terry, F., Gokemeijer, J., Mitra-Kaushik, S., Roberts, B. J., Tourdot, S., & De Groot, A. S. (2020). T-cell dependent immunogenicity of protein therapeutics preclinical assessment and mitigation—Updated consensus and review 2020. *Frontiers in Immunology, 11*, 1301. Available from https://doi.org/10.3389/fimmu.2020.01301.

Jin, J. F., Zhu, L. L., Chen, M., Xu, H. M., Wang, H. F., Feng, X. Q., Zhu, X. P., & Zhou, Q. (2015). The optimal choice of medication administration route regarding intravenous, intramuscular, and subcutaneous injection. *Patient Preference and Adherence, 9*, 923–942. Available from https://doi.org/10.2147/PPA.S87271.

Kessler, C., Edwards, E., Dissinger, E., Sye, S., Visich, T., & Grant, E. (2019). Usability and preference of epinephrine auto-injectors. *Annals of Allergy, Asthma & Immunology*, 256–262. Available from https://doi.org/10.1016/j.anai.2019.06.005.

Kirschbrown, W. P., Wynne, C., Kågedal, M., Wada, R., Li, H., Wang, B., Nijem, I., Badovinac Crnjevic, T., Gasser, H., Heeson, S., Eng-Wong, J., & Garg, A. (2018). Development of a subcutaneous fixed-dose combination of pertuzumab and trastuzumab: Results from the phase Ib dose-finding study. *Journal of Clinical Pharmacology, 59* (5), 702–716. Available from https://doi.org/10.1002/jcph.1362.

Krayukhina, E., Fukuhara, A., & Uchiyama, S. (2020). Assessment of the injection performance of a tapered needle for use in prefilled biopharmaceutical products. *Journal of Pharmaceutical Sciences, 109*, 515–523. Available from https://doi.org/10.1016/j.xphs.2019.10.033.

Lucas, A., Price, L., Schorzman, A., Storrie, M., Piscitelli, J., Razo, J., & Zamboni, W. (2018). Factors affecting the pharmacology of antibody-drug conjugates. *Antibodies, 7* (1), 10. Available from https://doi.org/10.3390/antib7010010.

Mager, D. E., & Jusko, W. J. (2001). General pharmacokinetic model for drugs exhibiting target-mediated drug disposition. *Journal of Pharmacokinetics and Pharmacodynamics, 28* (6), 507–532. Available from https://doi.org/10.1023/A:1014414520282.

Mao, C. P., Brovarney, M. R., Dabbagh, K., Birnböck, H. F., Richter, W. F., & Del Nagro, C. J. (2013). Subcutaneous vs intravenous administration of rituximab: Pharmacokinetics, CD20 target coverage and B-cell depletion in cynomolgus monkeys. *PLoS One, 8*(11), e80533. Available from https://doi.org/10.1371/journal.pone.0080533.

Markarian, J. (2019). Putting drug delivery into patients' hands. *Pharmaceutical Technology, 43*(6), 16–19. Available from http://www.pharmtech.com/putting-drug-delivery-patients-hands.

Mateos, M. V., Nahi, H., Legiec, W., Grosicki, S., Vorobyev, V., Spicka, I., Hungria, V., Korenkova, S., Bahlis, N., Flogegard, M., Bladé, J., Moreau, P., Kaiser, M., Iida, S., Laubach, J., Magen, H., Cavo, M., Hulin, C., White, D., & Usmani, S. Z. (2020). Subcutaneous vs intravenous daratumumab in patients with relapsed or refractory multiple myeloma (COLUMBA): A multicentre, open-label, non-inferiority, randomised, phase 3 trial. *The Lancet Haematology*, 7(5), e370–e380. Available from https://doi.org/10.1016/S2352-3026(20)30070-3.

McLennan, D. N., Porter, C. J. H., & Charman, S. A. (2005). Subcutaneous drug delivery and the role of the lymphatics. *Drug Discovery Today: Technologies*, *2*(1), 89–96. Available from https://doi.org/10.1016/j.ddtec.2005.05.006.

Nema, S., & Ludwig, J. D. (2019). *Parenteral Medications* (fourth ed.). Boca Raton: CRC Press, Taylor & Francis Group. Available from https://doi.org/10.1201/9780429201400.

Newman-Casey, P. A., Niziol, L. M., Mackenzie, C. K., Resnicow, K., Lee, P. P., Musch, D. C., & Heisler, M. (2018). Personalized behavior change program for glaucoma patients with poor adherence: A pilot interventional cohort study with a pre-post design. *Pilot and Feasibility Studies*, *4*(128). Available from https://doi.org/10.1186/s40814-018-0320-6.

Okegbe, T., Affo, J., Djihoun, F., Zannou, A., Hounyo, O., Ahounou, G., Bangbola, K. A., & Harris, N. (2019). Introduction of community-based provision of subcutaneous depot medroxyprogesterone acetate (DMPA-SC) in Benin: Programmatic results. *Global Health Science and Practice*, 7(2), 228–239. Available from https://doi.org/10.9745/GHSP-D-19-00002.

Patel, K., Parmar, S., Shah, S., Shore, T., Gergis, U., Mayer, S., & van Besien, K. (2016). Comparison of subcutaneous vs intravenous alemtuzumab for graft-vs-host disease prophylaxis with fludarabine/melphalan-based conditioning in matched unrelated donor allogeneic stem cell transplantation. *Biology of Blood and Marrow Transplantation*, *22*(3), 456–461. Available from https://doi.org/10.1016/j.bbmt.2015.10.022.

Rojas, L., Muñiz, S., Medina, L., Peña, J., Acevedo, F., Pinto, M. P., & Sanchez, C. (2020). Cost-minimization analysis of subcutaneous vs intravenous trastuzumab administration in Chilean patients with HER2-positive early breast cancer. *PLoS ONE*, *15*(2). Available from https://doi.org/10.1371/journal.pone.0227961.

Rombouts, M. D., Swart, E. L., Van Den Eertwegh, A. J. M., & Crul, M. (2020). Systematic review on infusion reactions to and infusion rate of monoclonal antibodies used in cancer treatment. *Anticancer Research*, *40*(3), 1201–1218. Available from https://doi.org/10.21873/anticanres.14062.

Sangave, N. A., Aungst, T. D., & Patel, D. K. (2019). Smart connected insulin pens, caps, and attachments: A review of the future of diabetes technology. *Diabetes Spectrum*, *32*(4), 378–384. Available from https://doi.org/10.2337/ds18-0069.

Scherer, K., Spoerl, D., & Bircher, A. (2010). Adverse drug reactions to biologics. *JDDG: Journal Der Deutschen Dermatologischen Gesellschaft*, *8*(6), 411–426. Available from https://doi.org/10.1111/j.1610-0387.2010.07339.x.

Scheuer, W., Friess, T., Burtscher, H., Bossenmaier, B., Endl, J., & Hasmann, M. (2009). Strongly enhanced antitumor activity of trastuzumab and pertuzumab combination treatment on HER2-positive human xenograft tumor models. *Cancer Research*, *69*(24), 9330–9336. Available from https://doi.org/10.1158/0008-5472.CAN-08-4597.

Shapiro, R. (2010). Subcutaneous immunoglobulin therapy by rapid push is preferred to infusion by pump: A retrospective analysis. *Journal of Clinical Immunology*, *30*, 301–307. Available from https://doi.org/10.1007/s10875-009-9352-2.

Shpilberg, O., & Jackisch, C. (2013). Subcutaneous administration of rituximab (MabThera) and trastuzumab (Herceptin) using hyaluronidase. *British Journal of Cancer*, *109*(6), 1556–1561. Available from https://doi.org/10.1038/bjc.2013.371.

Singh, S. K., Mahler, H. C., Hartman, C., & Stark, C. A. (2018). Are injection site reactions in monoclonal antibody therapies caused by polysorbate excipient degradants? *Journal of Pharmaceutical Sciences*, *107*(11), 2735–2741. Available from https://doi.org/10.1016/j.xphs.2018.07.016.

Stoner, K. L., Harder, H., Fallowfield, L. J., & Jenkins, V. A. (2014). Intravenous vs subcutaneous drug administration. Which do patients prefer? A systematic review. *Patient*, *8*, 145–153. Available from https://doi.org/10.1007/s40271-014-0075-y.

Taussky, P., Tawk, R. G., Daugherty, W. P., & Hanel, R. A. (2011). Medical therapy for ischemic stroke: Review of intravenous and intra-arterial treatment options. *World Neurosurgery*, *76*(6), S9–S15. Available from https://doi.org/10.1016/j.wneu.2011.05.048.

Tetteh, E. K., & Morris, S. (2014). Evaluating the administration costs of biologic drugs: Development of a cost algorithm. *Health Economics Review*, *4*(26). Available from https://doi.org/10.1186/s13561-014-0026-2.

The Electronic Medicines Compendium. (2021). https://www.medicines.org.uk/emc/.

Thomaidou, E., & Ramot, Y. (2019). Injection site reactions with the use of biological agents. *Dermatologic Therapy*, *32*(2). Available from https://doi.org/10.1111/dth.12817.

Tjalma, W., Huizing, M., & Papadimitriou, K. (2017). The smooth and bumpy road of trastuzumab administration: From intravenous (IV) in a hospital to subcutaneous (SC) at home. *Facts, Views & Vision in ObGyn*, *9*(1), 51–55. Available from https://www.ncbi.nlm.nih.gov/pmc/articles/PMC5506771/pdf/FVVinObGyn-9-51.pdf.

Tomar, D. S., Kumar, S., Singh, S. K., Goswami, S., & Li, L. (2016). Molecular basis of high viscosity in concentrated antibody solutions: Strategies for high concentration drug product development. *MAbs*, *8*(2), 216–228. Available from https://doi.org/10.1080/19420862.2015.1128606.

Tucker, C. A., Bebb, G., Klasa, R. J., Chhanabhai, M., Lestou, V., Horsman, D. E., Gascoyne, R. D., Wiestner, A., Masin, D., Bally, M., & Williams, M. E. (2006). Four human t(11;14)(q13;q32)-containing cell lines having classic and variant features of Mantle Cell Lymphoma. *Leukemia Research*, *30*(4), 449–457. Available from https://doi.org/10.1016/j.leukres.2005.08.016.

Turner, M. R., & Balu-Iyer, S. V. (2018). Challenges and opportunities for the subcutaneous delivery of therapeutic proteins. *Journal of Pharmaceutical Sciences*, *107*(5), 1247–1260. Available from https://doi.org/10.1016/j.xphs.2018.01.007.

Usmani, S. Z., Nahi, H., Mateos, M. V., van de Donk, N. W. C. J., Chari, A., Kaufman, J. L., Moreau, P., Oriol, A., Plesner, T., Benboubker, L., Hellemans, P., Masterson, T., Clemens, P. L., Luo, M., Liu, K., & San-Miguel, J. (2019). Subcutaneous delivery of daratumumab in relapsed or refractory multiple myeloma. *Blood*, *134*(8), 668–677. Available from https://doi.org/10.1182/blood.2019000667.

Vermeire, S., D'heygere, F., Nakad, A., Franchimont, D., Fontaine, F., Louis, E., Hootegem, P. V., Dewit, O., Lambrecht, G., Strubbe, B., & Baert, F. (2018). Preference for a prefilled syringe or an autoinjection device for delivering golimumab in patients with moderate-to-severe ulcerative colitis: A randomized crossover study. *Patient Preference and Adherence*, *12*, 1193–1202. Available from https://doi.org/10.2147/PPA.S154181.

Wallace, Z., Harkness, T., Blumenthal, K., Choi, H., Stone, J., & Walensky, R. (2020). Increasing operational capacity and reducing costs of rituximab administration: A costing analysis. *ACR Open Rheumatology*, *2*(5), 261–268. Available from https://doi.org/10.1002/acr2.11133.

Wright, J., & Jones, G. (2017). Developing the subcutaneous drug delivery route. *Medical Research Archives*, *5*(12). Available from https://esmed.org/MRA/mra/article/view/1652.

CHAPTER 6

Formulation and device lifecycle management of biotherapeutics—decision architecture

Introduction

To enable timely availability of the most convenient drug delivery modality, manufactures need to consider early during a biotherapeutical's lifecycle whether and when to embark into respective formulation and device lifecycle management. At the different stages of a molecule's lifecycle, that is, from early discovery up to its first introduction into the market and beyond, emerging insights into product optimization needs and opportunities are becoming available. As a consequence, early engagement into formulation and device lifecycle management can be challenging due to an initially limited understanding of molecule properties, customer requirements, and market dynamics that can guide respective developments. To overcome this hurdle, in addition to leveraging experience from other molecules applying similar drug delivery changes (refer to Chapter 5, Formulation and device lifecycle management of biotherapeutics—leveraging synergies across therapeutic areas, for details), scientists need to foresee and plan related assessments and studies to allow informed decision-making.

The focus of this chapter is on the rationale and practical conduct of formulation and device lifecycle management for biopharmaceuticals, such as monoclonal antibodies (mAbs). In scope are drug delivery improvements that advance the product presentation, reduce dosing complexity, and, in case the biological has been demonstrated safe and tolerated, enable dosing outside of a healthcare institution.

In the simplest case, and today this is broadly happening as a routine product optimization, drug delivery scientists and engineers aim at extending the handling times and shelf-life of a medication, developing ready-to-use liquid dosing solutions instead of powders for reconstitution, and implementing standardized packaging compatible to commonly used instruments. More recently, focus of formulation and device lifecycle management

Formulation and Device Lifecycle Management of Biotherapeutics
DOI: https://doi.org/10.1016/B978-0-12-823741-0.00003-8
© 2022 Elsevier Inc.
All rights reserved.

is on reducing the treatment burden for patients and the healthcare system overall. Related activities include fast intravenous infusion regimens, alternative and less invasive administration routes, and fixed-dose regimens replacing body size-adjusted dosing schedules. Connected drug delivery devices are investigated with the objective to support drug administration outside of a healthcare institution. For biotherapeuticals that are a component of combination therapy, the inherent dosing complexity can be further reduced by aligning the drug administration frequency with that of a combination partner, as well as with the availability of fixed-dose combinations or dual-chamber devices capable of administering two or more molecules via one subcutaneous injection (refer to Chapter 1, Enabling customer-centric and sustainable drug delivery—formulation and device lifecycle management of biotherapeutics).

Why should manufacturers embark into formulation and device lifecycle management?

The underlying reasons for optimizing a biopharmaceutical's delivery profile and product presentation during its lifecycle can be divided into four main pillars (Fig. 6.1). Namely, changes in drug delivery have the potential (1) to improve the efficacy and or safety profile, (2) to increase the efficiency in disease management, (3) to broaden access to medications, and (4) to extent exclusivity of the medication.

Formulation and device lifecycle management to improve efficacy and/or safety of a biotherapeutic

In some instances, product optimizations improve not only patient's experience with the drug administration procedure but also efficacy and/or

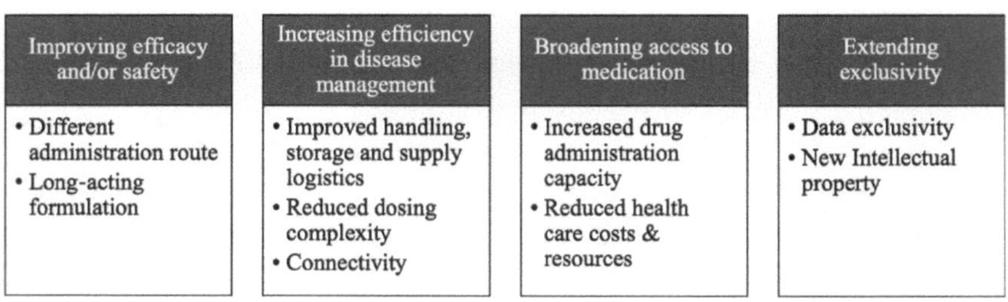

Figure 6.1 Reasons and ways to optimize a biopharmaceutical's delivery profile during its lifecycle.

safety of a biotherapeutical. Underlying mechanisms include targeting delivery of a molecule directly to its site of action in a tissue or organ (Singh et al., 2019) or changing the release profile into the systemic circulation with a different administration route or formulation technology.

Targeted drug delivery technologies that carry a molecule directly to its site of action alter the biodistribution profile of a biotherapeutical as compared to conventional formulations and thus still require the conduct of larger clinical trials to address any associated changes in the safety and efficacy profile in humans.

Drug delivery strategies that modify the release profile of a biotherapeutical molecule into the systemic circulation include changes in the administration route (i.e., subcutaneous instead of intravenous dosing) and subcutaneous or intramuscular long-acting or sustained release formulations. In theory, as a consequence of the slower absorption into the systemic circulation as compared to intravenous administration, such modifications in the administration profile may reduce side effects related to high maximum serum levels (C_{max}). Especially, mAbs that have been reported to exhibit severe infusion-related reactions during the initial intravenous infusions may benefit from a reduced C_{max}. For daratumumab, approved for the treatment of multiple myeloma, a lower rate of infusion-related reactions was observed following subcutaneous compared to intravenous administration (Paul et al., 2020). This observation is in line with the lower and later peak concentrations inherent to subcutaneous injection, while higher C_{max} values are measured directly after intravenous infusion (Clemens et al., 2018). Taking into account these observed benefits of a reduced C_{max}, it becomes evident that further lowering and delaying C_{max} could even more reduce related injection reactions. In this context, injectable depots represent an interesting drug delivery alternative. These vehicles are prepared by encapsulation of the drug in, for example, biodegradable, injectable depot formulations facilitating long-term controlled drug release. Poly(lactic-co-glycolic acid)-based micro- or nanoparticulate systems or implants are technologies that are already applied to several small molecules and peptides (Schwendeman et al., 2014).

Formulation and device lifecycle management to increase the efficiency in disease management

Another key aspect for conducting formulation and device lifecycle management for biotherapeuticals is to generally increase the efficiency in disease management via an improved drug delivery profile or product presentation. This includes storage, handling, and shipment parameters that also can become important factors in the management of clinical supplies. As an example, an extended product handling time may allow rescheduling an appointment in case of a short notice cancellation and using the prepared medication for the next consultation. Product presentations not requiring temperature controlled shipment or cool storage result in lower supply chain and inventory costs and can be more efficiently distributed and stored (Sykes, 2018). Another important feature required to be incorporated into the product presentation are anticounterfeit measures including serialization to allow patients and pharmacists fast and reliable confirmation of the identity and validity of the medication prior to usage (Bansal et al., 2013).

Reducing the dosing complexity with fixed dosing regimens, less-invasive and faster administration modes, and ready-to-use drug administration devices has the potential to cut down drug delivery related resources in the clinic and to even shift treatment to the home setting. Complemented with connected injection devices and accompanying health apps, such product optimizations can facilitate drug application in the home setting, while still allowing for an ongoing dialog between the patient and healthcare provider (Bittner et al., 2019). In the future, electronically-derived data on patient adherence behavior may have the potential to complement cost-effectiveness analyses and influencing drug approvals (Tjalma et al., 2018).

Formulation and device lifecycle management to broaden access to medications

Inherent to the reduced dosing complexity, a more cost effective drug delivery profile and product presentation can ultimately broaden access to medication, the third pillar rationalizing formulation and device lifecycle management. For instance, fixed-dose subcutaneous dosing alternatives to intravenous infusions have been demonstrated to decrease drug administration burden in the clinic (Rule et al., 2014; Tjalma et al., 2018) and thus can address infusion capacity bottlenecks.

Optimization of packaging to allow for shipment and storage at ambient conditions simplifies logistics and is especially relevant in countries without a reliable infrastructure for low temperature cool chain transport and adequate storage capacities (Shah et al., 2021). For thermosensitive biotherapeuticals requiring controlled refrigeration, intelligent packaging

(Hui et al., 2020) including temperature monitoring can help to ensure the quality of the medication.

Drug delivery alternatives and optimized product presentations that enable treatment outside of a controlled environment will not only increase capacity in the clinic but also improve medication access for patients living in rural regions far from a healthcare institution (Gardulf, 2007; Sehn et al., 2007). Here, electronic patient diaries can provide the responsible physician with a better understanding of whether a patient follows a therapeutic medication regimen and allows for more personalized prescribing practices (Klonoff et al., 2020). For example, a smart insulin pen that includes a sensor collecting and storing data on the date and time of injections and the number of units administered can share information with the healthcare provider and thus provides an accurate picture of the relationship between actually administered doses of insulin and levels of glycemia (Klonoff & Kerr, 2018).

Formulation and device lifecycle management to extend exclusivity

There are instances in which product optimizations adopting novel formulation or device technologies can result in additional intellectual property (Beall & Kesselheim, 2018; Gupta et al., 2010). This is under the presumption that applying a formulation or device technology either bears a tangible impact on efficacy and/or safety of a biotherapeutical, on patient experience, satisfaction, and compliance with the treatment schedule, or if a marked societal or health economic impact could be demonstrated. In addition, in the event that the technology development was supported by significant clinical trials, regulators can provide so-called data exclusivity as incentive to compensate the innovator company for the investment (FDA, 2016). During the period of data exclusivity, preclinical and clinical data generated by the manufacturer cannot be referenced in the regulatory filings of another company for the same drug substance (Singh et al., 2017).

When to embark into formulation and device lifecycle management—general considerations

Chapter 1 introduced the different aspects of formulation and device lifecycle management aiming at reducing dosing complexity and providing patient-centric and sustainable product presentations. Focus is on technologies for biopharmaceuticals that (1) improve the logistics around supply chain, as well as drug preparation and administration in a healthcare institution, (2) reduce the treatment burden via decreased dosing complexity and time expenditure in a healthcare institution, and possibly enable a flexible care setting, and (3) minimize dosing complexity inherent to combination therapy with two or more molecules.

Along the lifecycle of a medicine, manufacturers increasingly gain a more robust understanding of product optimization needs and opportunities. Emerging insights are derived from the results of nonclinical and clinical studies, customer feedback, as well as from changes in the environmental landscape and need to be assessed systematically and early on. The following sections provide a decision architecture on when manufacturers are able to decide on the most appropriate product optimization based on the increasing knowledge that becomes available during the maturing lifecycle of a biotherapeutical.

Nonclinical development

The nonclinical development stage encompasses all experiments conducted prior to entering into clinical trials. Efforts are made to determine whether a novel biological molecule has the potential to efficiently act at its target site in the body at a reasonable dose level. The key focus is on demonstrating the novel molecule's mode of action in preclinical models. At this stage development, efforts are driven predominantly by nonclinical pharmacokineticists, toxicologists, and pharmacologists, as well as by clinical pharmacologists who base the design of phase 1 clinical studies on the outcome of the nonclinical work.

Aspects of improving drug delivery typically are of secondary priority during this early development phase as information relevant for product optimizations is still very limited. However, nonclinical studies should be designed to inform on the feasibility and the most appropriate conduct and timing of possible product optimizations and thus include assessments on alternative drug administration modalities. Without a significant increase in costs and resources, studies that are anyway required to characterize the novel biological molecule can include additional arms to test different administration routes.

A prerequisite for delivering a biological molecule with a route different to intravenous infusion is that its relative bioavailability is high enough to achieve the target blood concentration at an equitable dose level.

Supporting data can be generated either in dedicated nonclinical pharmacokinetic models or as part of toxicokinetic assessments in tolerability and safety trials. Research is ongoing to establish a correlation between the bioavailability in different animal models and the bioavailability in man (Richter & Jacobsen, 2014). It is expected that with the growing database in the field, over time, increasingly accurate predictions of the administration route-dependent bioavailability in humans will be realized.

In addition to the relative bioavailability of a biological following different administration routes, toxicokinetic studies establish a correlation between pharmacokinetic parameters and the safety and tolerability profile of a biological substance. For example, in the event that data reveal a correlation between C_{max} and specific adverse events, a subcutaneous formulation might be preferred over intravenous administration due to the intrinsically lower maximum concentrations following injection into the interstitial space. On the contrary, in case subcutaneous administration of a biopharmaceutical results in local tolerance issues, intravenous dosing might be favored.

Likewise, combined nonclinical pharmacokinetic and pharmacology studies can demonstrate a possible correlation between pharmacokinetic parameters and pharmacodynamic endpoints and give an early direction on whether the differences in pharmacokinetic parameters with different administration routes are expected to impact clinical outcomes. For instance, in case a biological exhibits a relatively short intrinsic half-life, subcutaneous, or intramuscular slow release formulations (Rahnfeld & Luciani, 2020; Schaefer et al., 2016) that expand the residence time in the body should be tested for technical feasibility early on.

The desired dosing frequency of a novel biological molecule in a given indication has to be determined with the overall disease management schedule in mind. For example, the drug administration schedule should ideally be aligned with the number of regular clinical visits for physical examinations . Especially, if the molecule will have to be administered in the clinic, aligning routine visits with dosing days will reduce the overall treatment burden for patients and the healthcare institution. In this context, any information on the expected dosing frequency of possible combination partners that are either already on the market or in late stage development will further inform the design of nonclinical studies and subsequent clinical trials. Even in the absence of information on future disease management, applying early pharmacokinetic/pharmacodynamic modeling to explore possible dosing frequencies for a range of dose levels could be highly valuable. Under the assumption that less frequent parenteral administration is preferred by patients and healthcare providers, such predictions can support anticipation of acceptable dosing regimens.

Timely engagement of medical, commercial, and market access experts with a sound understanding of the target indication is therefore critical to gain insights into the current and more importantly the expected future environment in the disease area. Knowledge on how molecules are dosed, on patient preferences and capabilities, as well as on how the dosing schedule of future biologics will possibly change the present standard of care is mandated to decide on the necessary elements of formulation and device lifecycle management.

During nonclinical development of a biological molecule, drug delivery and formulation scientists play a pivotal role in front-loading activities for formulation development. A variety of technical formulation and device assessments can begin in parallel to nonclinical in vitro and in vivo research. As the final dose level of a biological molecule is typically not yet known at this stage, it is recommended to test a range of technical options to avoid starting from scratch following readouts from clinical studies. Especially for subcutaneous administration and with the aim to reduce the overall dosing volume in the clinic, the feasibility of a high concentration solution should include dosing solutions of different concentrations and with different excipients. Samples can be put on long-term stability at different temperatures, storage conditions, and in primary containers of different compositions. In addition, in case a potential future biological combination partner is already suggested based on scientific plausibility, stability studies that support coadministration and possibly fixed-dose combination development can be initiated. This approach ensures evaluation of a broader range of formulation and packaging opportunities and increases the likelihood that a robust product configuration can later proceed into clinical trials.

Likewise, the need for a drug delivery device can already be anticipated early on. In the event that drug administration outside of a healthcare institution is already standard of care in the target indication and if the patient population is expected to be capable of self-administration, a ready-to-use device should be foreseen from the beginning. Indubitably, in the absence of the final dose decision and the expected dosing volume, selecting a device type is challenging. However, stability of the molecule in different primary device container materials and at different concentrations will inform device type selection later on.

Phase 1 clinical development

In phase 1 clinical studies, a new biological molecule is administered to healthy volunteers or patients, respectively (Mahan, 2014). Trials are conducted by clinical pharmacologists in close cooperation with medical, biomarker, and safety scientists. Typically, in these first-in-human trials, the dose levels tested are slowly increased with the aim to identify the dosage that is expected to be efficacious and safe based on the preceding animal studies. Key focus is on the pharmacokinetic profile, including the absorption rate, the bioavailability, and the distribution and elimination pattern, as well as on the safety and tolerability of the novel biotherapeutic.

Both, a molecule's pharmacokinetics and safety, can be affected by a change in the way it is delivered to the body, that is, with a different administration route or dosing frequency, or by using different formulation ingredients. Related assessments should consequently be included in phase 1 studies early on. Practically, for molecules for which the administration route has not yet been decided, clinical scientists may plan to conduct the initial dose escalation part via the intravenous route and to add an additional adaptive cohort with subcutaneous administration once the target dose has been reached. This approach limits the number of subjects required for the clinical trial and allows comparing the pharmacokinetic profile including the relative bioavailability of the different administration routes at one dose level only (Fig. 6.2).

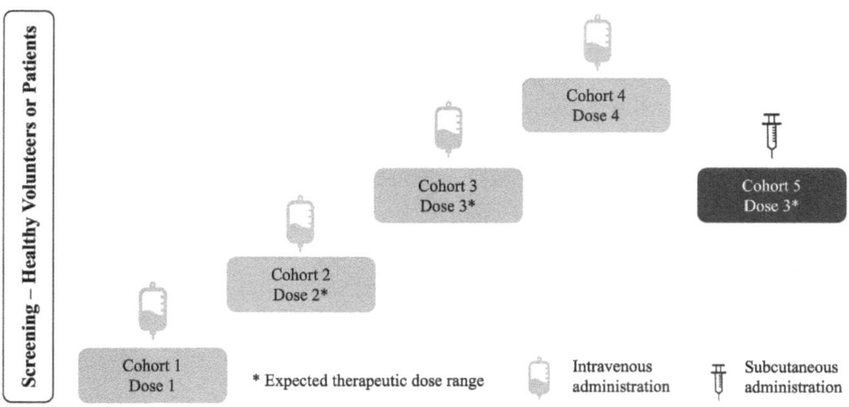

Figure 6.2 Possible design of phase 1 trial with intravenous dosing of a novel biotherapeutical including a subcutaneous cohort extension. Assessing both intravenous and subcutaneous dosing in the first-in-man studies for a novel biological molecule reduces the patient numbers and costs required for two separate studies and enables early decision-making on the most appropriate administration route(s) to be brought forward into phase 2 and/or 3 clinical trials. *Abbreviations: IV, intravenous; SC, subcutaneous.*

Today it is acknowledged that allowing for variation in the dosing procedure already in phase 1 clinical trials will help generate data that support future formulation and device lifecycle management of a molecule (Bittner et al., 2012; Bruin et al., 2020). Flexibility in the dosing procedure accounts for local practice and therefore reflects the variety of expected drug administration conditions the real-world setting. This is particularly important when considering an impact of deviations in drug administration practice on the pharmacokinetics of a molecule (Berger, 2004). Drug delivery parameters in scope for a parenterally dosed biopharmaceutical would include the infusion rate and the nature of the infusion equipment, such as syringe pumps and associated infusion systems. Here, early clinical data on a broader range of dosing conditions will support future device developments, as the device type-dependent injection rates would be bracketed by the observed rates in the studies.

Phase 1 data also provide insights into the impact of body size on the biodistribution of a molecule. These data are used to build the pharmacokinetic model that is applied to assess whether fixed doses or fixed-dose bands for defined weight categories instead of body weight- or body surface area-adjusted dosing might be feasible. Fixed-dose regimens have the potential to reduce drug wastage and the overall dosing complexity (Chatelut et al., 2012) and are a prerequisite for developing ready-to-use subcutaneous injection devices that can be applied by a patient without individual dose adjustment steps.

In case a combination therapy partner for a novel biotherapeutic has already been identified prior to or during phase 1 clinical development, codosing trials assessing the reciprocal impact on each other's pharmacokinetic and safety profiles might be considered. While more meaningful data on the opportunities and challenges of the combination will become available in phase 2 or 3 clinical trials, information from first-in-human studies on the tolerated dose levels with the combination partners supports an initial assessment of the expected dosing volume and thus the preselection of an injection device. Moreover, pharmacokinetic-based modeling approaches can complement the available database in determining the dosing regimen and frequency and in predicting whether these can be aligned with possible combination partners.

Especially if phase 1 clinical trials are conducted in patients, adding preference and satisfaction questionnaires around different drug delivery aspects can greatly support the design of a customer-centric product presentation and dosing frequency. Technical studies on the stability of the formulation that continue to run in parallel to phase 1 studies allow for a more precise estimate of the achievable concentration of a subcutaneous dosing solution that directs the future dosing volume. Taken all of this information together, manufacturers will have solid insights on whether a certain drug delivery profile or administration route is tolerated, sufficiently bioavailable and can be administered with a reasonable dosing volume. Available insights can guide the decision on pursuing one over another drug delivery option already at this development stage; that is, in the absence of efficacy readouts from phase 2 and/or 3 clinical trials.

Phase 2 and 3 clinical development

Phase 2 clinical investigations are conducted to determine the dose of a novel biotherapeutic molecule based on efficacy and safety endpoints that will subsequently be confirmed in larger phase 3 studies. These trials can either be conducted sequentially or combined within one single adaptive study (Chow & Chang, 2008; Lu et al., 2020). Here, the biological compound is compared to either the current standard of care in the indication or to a placebo, in case a standard therapy has not yet been established. At this stage of development, efforts are predominantly driven by medical and clinical experts in the disease area, as well as by clinical operations specialists who are responsible for the conduct of the trials within the preplanned budget and timelines.

Changes in drug delivery modalities following completion of these pivotal studies that could impact the molecule's pharmacokinetic and or safety profile would require conducting additional phase 1 study(ies) at a minimum. Consequently, if timely feasible, the final formulation, device, and dosing regimen should be tested during phase 3 studies to provide the desired product presentation already at the initial molecule approval. If this cannot be realized due to either an insufficient database or in case a product optimization would cause a delay in the approval process, respective development activities would need to be conducted as a postapproval lifecycle management activity using a so-called "clinical bridging approach" (Bittner, et al., 2012; Bruin et al., 2020) (for more details please see also "Postapproval product optimizations").

Following phase 2 and 3 clinical trials, a larger database is becoming available that further reduces the uncertainty around the most appropriate drug delivery profile. For instance, after phase 1, the feasibility of a fixed dosing regimen was based on the pharmacokinetic, safety, and if already available, exploratory efficacy data from the first-in-human studies only. Now, the database is complemented with the available and statistically powered efficacy and safety data in the target patient population(s). These data further validate the clinical appropriateness of the selected dosing regimen.

Even if the final drug delivery device had not yet been studied in phase 2 or 3 trials, the clinical data generated in the pivotal studies together with technical feasibility assessments that are running in parallel, generally suffice to select the most appropriate device platform. Forthwith, the final dose, the feasible concentration of the dosing solution and, ultimately, the dose volume, are known. Moreover, provided the drug administration conditions, such as the infusion rate or the dosing equipment, did allow for a comparatively broad variability in the drug administration parameters, the available data pool will allow for defining the appropriate specifications of the device.

If applicable, manufacturers may include a study arm assessing dosing outside of a healthcare institution to inform about associated challenges that could be addressed in educational materials, that is, via connected devices and accompanying health apps. Preference and patient satisfaction questionnaires will confirm the need for a device and device type to enable treatment in a flexible care setting. Health economic assessments, exploring the impact of differences in the dosing regimen on healthcare costs, charges, and expenditures, will generate insights into the broader societal value of the most desirable drug delivery profile and product presentation.

During phase 2 and 3 clinical trials, a number of patients will already have received the novel biological together with a combination therapy partner and additional information on emerging new cotreatments is becoming available. Related insights will further inform a possible development of fixed-dose combinations or dual-chamber devices.

Postapproval product optimizations

Following market entry of a novel biological therapy, real-world evidence on its usage in the disease and its efficacy and safety profile is evolving, adding to the insights generated in the controlled clinical trial setting. Still, ongoing product optimizations remain an integral and important part of the biotherapeutic's lifecycle management. Consequently, even when approved, cross-disciplinary cooperation is required and respective product teams include experts from clinical sciences and operations, technical development, supply chain management, commercial and access management, and regulatory and medical affairs.

At this stage, awareness of possible drawbacks of the product configuration and dosing regimen is becoming transparent. Challenges related to storage, shipment, or handling of the initial product presentation can mandate changes in the packaging configuration and the manufacturer may have received consistent feedback from patients and healthcare providers on difficulties with the dosing procedure.

Embarking into the development of novel drug delivery technologies with limited proof-of-concept during a development phase in which the efficacy and safety profile of a novel biotherapeutic has not yet been established bears a considerable risk of failure. This is especially important in case the underlying development activities for the new technology could delay initial approval of the molecule. Once a molecule is approved and the treatment concept has been demonstrated, investment into drug delivery improvements is performed with an overall increased probability of success.

In the event that emerging healthcare requirements command a specific product optimization for a product that is already on the market, respective development activities are being initiated. Related insights include feedback from patients, patient organizations, or healthcare providers for instance on the need for an improved dosing regimen to ensure adherence and compliance with the treatment procedure. Also, aspects of drug delivery that reduce dosing complexity and provide the option to shift treatment outside of a healthcare institution can become more important over time. Currently, the demand for formulation and device lifecycle management that reduces dosing complexity and thus enables a flexible care setting has drastically increased in light of the growing pressure on the healthcare systems and in particular due to the COVID-19 pandemic.

In this context an important consideration when evaluating the value of drug delivery for the healthcare system in a given legislation is that the present policies for reimbursing drug administration may change with the availability of novel dosing alternatives. In a situation where biotherapeutics are administered via more complex and time-consuming intravenous infusions, healthcare provider reimbursement models, such as the fee-for-service model (Quinn et al., 2020), are typically valued by physicians to compensate for their efforts related to drug administration. With the availability of ready-to-use subcutaneous dosing alternatives that reduce the burden associated with dose calculation and delivery, reimbursement models that value aspects of efficient dosing will increasingly gain importance (Vlaanderen et al., 2019).

Once a novel biological treatment is approved and used in routine clinical practice, a rich database that is relevant to assess any further demand for improvements in aspects of drug delivery is gradually becoming available. This includes results from long-term extensions of the pivotal phase 3 studies, postapproval commitment trials that investigate safety and tolerability aspects in a larger population, as well as clinical studies examining new indications and possible line extensions. This could include assessments of future combination partners, a different administration route, or a different dosing frequency.

For example, the initial label for a novel biotherapeutic may have mandated a considerably slow intravenous infusion rate to reduce infusion related reactions (Cáceres et al., 2019). The growing evidence from real-world data on the frequency of these events may reveal that infusion related reactions are rare in later treatment cycles (Rombouts et al., 2020) and management of an infusion reaction could be individualized based on patient symptoms and status (Vogel, 2010). In this instance, the manufacturer may pursue a data-based label update to allow for a faster infusion for eligible patients following treatment initiation at slower infusion rates. Alternatively, a dosing regimen could have been reported to result in a significant number of errors due to body size-adjusted dosing or due to a complicated administration procedure (Sabzi et al., 2019). Here, the development of a fixed dose or a dose band-based dosing regimen and/or of a ready-to-use drug delivery device may be considered to improve the overall product profile. Also, in case a biotherapeutic is approved for home- and self-administration, but adherence to treatment in this non-controlled setting has been reported to be a consistent issue, development of a connected device with accompanying health app that includes dosing reminders, adherence trackers, and patient diaries and enables a remote dialog between patient and healthcare providers could be considered. Regarding the development of a fixed-dose combination or dual-chamber devices that facilitate coadministration, more information on emerging new cotreatments is becoming available (Bittner et al., 2018).

Nonclinical and clinical development approaches bridging from established to novel dosing regimens that are initiated following molecule approval, typically, do not require large-scale clinical trials. A comparatively large efficacy and safety database, including insights into a possible correlation with pharmacokinetic and biomarker data, has become available. Here, a demonstrated pharmacokinetic–pharmacodynamic correlation serves as a basis for the design of smaller nonclinical and clinical trials. Focused technology developments are further informed by input from patients and healthcare providers on current product features that can pose a challenge for drug administration, especially, when it comes to dosing outside of a controlled setting.

Simple product optimizations, such as a change in the formulation excipients to extent the shelf-life and improve storage conditions, or the application of novel drug administration devices are not expected to markedly the efficacy and safety profile of a molecule. In such cases manufacturers are typically required to only generate supportive pharmacokinetic data ranging from relative bioavailability studies to demonstration of pharmacokinetic equivalence, that is, bioequivalence, either in healthy volunteers or patients (Hu et al., 2020; Shin et al., 2020).

The nonclinical and clinical bridging program will be larger for a change in route of administration, for example, intravenous to subcutaneous dosing. Such a switch in administration route impacts the absorption into the systemic circulation and thus alters absorption rate and maximum exposure (C_{max}) of the molecule. Here, a nonclinical bridging program could consist of pharmacokinetic, pharmacodynamic, and toxicology studies to identify possible differences in the biodistribution, efficacy surrogates, and tolerability. Clinical phase 1 trials would subsequently confirm the pharmacokinetic and safety profile in man. In the event that C_{max} had been shown previously not to contribute to efficacy and safety, a phase 3 trial to compare the novel subcutaneous dosing regimen with the previous intravenous product presentation would focus on pharmacokinetics, safety, and specifically for biological molecules, immunogenicity has to be assessed as well. Efficacy endpoints would be added as supporting information (Shpilberg & Jackisch, 2013). The underlying concepts and the conduct of related nonclinical and clinical bridging programs are explained in Chapter 2, clinical development of subcutaneous dosing alternatives to established intravenous formulations for monoclonal antibodies—pharmacokinetic-based bridging approach.

When to embark into formulation and device lifecycle management—subcutaneous dosing alternative as a case study

The above sections provided a general description on product optimization aspects that should be addressed during the different stages of a biotherapeutical's lifecycle. In the following, practical recommendations are made on when such activities could start depending on the available technical, nonclinical, and clinical database for a novel biologic. For biological molecules in scope, C_{max} does not contribute to clinical efficacy and therefore intravenous and subcutaneous formulations are considered interchangeable when administered with the same dosing frequency at a dose that results in comparable serum trough concentrations (C_{trough}) and area under the serum concentration−time curve. Chapters 2 and 3 illustrated the scientific basis for the nonclinical and clinical development pathways for subcutaneous dosing and injection device alternatives based on prior examples.

Introduction of case study

In the hypothetical case study, a manufacturer is developing a novel mAb and would like to offer a subcutaneous formulation, ideally delivered with a ready-to-use injection device, thus allowing drug administration in a flexible care setting, including the option of home- and patient self-administration.

To realize this objective, a number of questions need to be answered throughout the molecule's lifecycle. Key aspects in this context are (1) the technical and clinical feasibility of subcutaneous dosing, (2) the selection of an appropriate injection device, and (3) the suitability of the molecule for administration outside of a healthcare institution. Table 6.1 illustrates the emerging insights on the technical and clinical feasibility of this target product profile that become available over time.

Decision architecture

The feasibility of a subcutaneous formulation is depending on the stability of a high-concentration formulation required to minimize the injection volume, as well as on the molecule- and formulation-dependent local tolerability profile. Initial short-term stability data with prototype subcutaneous formulations can be generated early on. Even during nonclinical and phase 1 clinical development, that is, prior to confirmation of the final dose,

Table 6.1 Development of a subcutaneous dosing alternative for biotherapeuticals.

Parameter	Data requirements	Available information following completion of different development stages			
		Nonclinical development	Phase 1 clinical development	Phase 2/3 clinical development	Market
Subcutaneous formulation	Stability of high-concentration formulation	+	++	+++	+++
	Tolerability profile	–	+	+++	+++
Injection device	Dose	–	++	+++	+++
	Dosing volume	–	++	+++	+++
	Fixed dose or fixed-dose bands	–	++	+++	+++
	Compatibility of formulation with primary container	+	++	+++	+++
Flexible care setting	Systemic safety profile	–	+	+++	+++

Note: Emerging information to support decision-making throughout the molecules lifecycle.

manufacturers may decide to pursue a so-called bracketing approach, whereby formulations that account for different dose levels are put on stability. While the composition of the formulation may not be final at this stage, manufacturers can take initial decisions on pursuing one over the other administration route based on this early technical information. Following phase 1 and 2 clinical studies, the dose to be confirmed in phase 3 trials has been selected. In case changes to the initial formulation composition are mandated, at that time, the clinical formulation can be entered into stability testing. Studies on the local tolerability following subcutaneous injection would be part of nonclinical tolerability studies first, followed by analogous assessments in phase 1 clinical trials. Provided these studies are conducted in the target patient population, related tolerability data can already provide initial confidence in the local tolerability profile with subcutaneous injections. Phase 3 studies will confirm the acceptability of local reactions with the final dose level in the final dosing vehicle.

The choice of a suitable injection device is depending on the dose, the dosing volume, the feasibility of fixed doses or fixed dose band, as well as on the compatibility of the formulation with different primary container materials. As described in Chapter 3, in the absence of a device technology platform, clinical studies are frequently conducted with a prefilled syringe (PFS) for lower subcutaneous dosing volumes and with a vial containing the dosing solution that is either injected manually using a handheld syringe or an infusion set (IS) for higher volumes. Autoinjectors (AIs), pen devices, or wearable on-body delivery systems (OBDSs) are subsequently developed as a lifecycle management activity.

Clinical information on the expected dose range and related dosing volumes is becoming available following phase 1 clinical studies. The pharmacokinetic and safety data and, if the study is conducted in patients, exploratory efficacy readouts are used to calculate the doses to be further pursued in clinical phase 2 and 3 trials. At this stage, manufacturers are in a position to preselect suitable injection device technologies. To unfold the full value of a ready-to-use device, ideally, the biological molecule is administered with a fixed dose or fixed-dose bands, so that only a limited number of dose strengths have to be produced. In case phase 1 studies reveal that this is not feasible, devices that allow for adjusting the individual dose volume via a dial dosing mechanism might be developed (Pisano, 2014), however, with the drawbacks of possible dosing errors and an increased amount of drug wastage. It is recommended that as soon as the need for an injection device is known, that is, even prior to phase 1 trials, batches with different doses and formulations are put on stability with various primary container materials. This way, decisions on the most appropriate device type can be frontloaded to the extent possible. Longer-term stability studies would subsequently confirm feasibility of the final dose, dosing volume, and formulation configuration.

The feasibility of dosing in a flexible care setting is primarily depending on the systemic safety profile of the biotherapeutic agent. Associated data are available following completion of the pivotal phase 3 studies. Especially hypersensitivity and anaphylactic reactions are of relevance in this context. Such reactions that occur predominantly during the first injection cycle(s) can be of different severity for individual patients and will guide regulators and the responsible physician on when and under which circumstances to allow home- and self-administration for eligible molecules (refer to Chapter 4, Enabling a flexible care setting in oncology—how to realize home- and self-administration of biotherapeutics in cancer care for detailed background information).

Development scenarios for a subcutaneous dosing regimen

Rationale for pursuing different development scenarios

Traditionally, many mAbs were first developed with an intravenous formulation, while subcutaneous development was initiated as a subsequent lifecycle management. The reasons behind this stepwise approach are manifold and depend on the physicochemical and clinical characteristics of the individual mAb, the dose level and associated dosing volume, as well as on the current and expected future standard of care in the respective disease area.

When starting development of a novel biotherapeutic molecule, manufacturers need to assess whether the molecule is suitable for subcutaneous administration based on the feasibility of a stable high-concentration dosing solution and the tolerability following subcutaneous injection. In case underlying formulation development would markedly delay initial launch of the biologic, an intravenous dosing alternative would be preferred as first-to-market product presentation.

Especially high-dose mAbs with corresponding high-volume dosing solutions did pose challenges for manufacturers. To avoid exceeding individual injection volumes of 1–2 mL, subcutaneously administered treatments were given more frequently as compared to their intravenous counterparts, thus diminishing to some extent the convenience benefit of the less invasive route. Today, these challenges have been overcome to a great extent with the feasibility of high-concentration formulations, the coformulation of a dispersion enhancer, or the development of wearable high-volume OBDSs (refer to Chapter 1, Enabling customer-centric and sustainable drug delivery—formulation and device lifecycle management of biotherapeutics).

Another aspect that drives the timing of developing a subcutaneous formulation is the treatment setting in the respective disease area for a novel mAb. In settings in which subcutaneous administration is already the standard-of-care, novel biotherapeutics would be introduced with this administration route at initial launch. The underlying reasons are that acceptability of an intravenous dosing regimen with the need to receive the medication in a healthcare institution would typically not be preferred by patients used to home- or even self-administration, unless the molecule has a proven efficacy benefit over established medications. Moreover, the inherently high administration costs and resources would be a drawback and constitute an increased burden on the healthcare system. In areas where intravenous administration is still standard and combination therapy partners are primarily dosed intravenously, having an intravenous formulation at initial molecule launch would account for hospital and pharmacy practices and thus simplify drug delivery overall. A subcutaneous dosing alternative could follow sequentially to account for a changing drug delivery landscape and the option for enabling a flexible care setting.

Practical conduct of different development scenarios

Generally, a manufacturer may decide to pursue mAb development with only one administration route, that is, intravenous or subcutaneous. In this case, the basic development program does not differ, except for particularities related to the feasibility of the injection. The program would start with technical formulation development and the nonclinical toxicology and pharmacology studies followed by the clinical development program from phase 1 up until phase 3 investigations.

Considering the above reflections on the value of the most appropriate launch sequence and the respective availability of the intravenous and subcutaneous formulations, the development of a subcutaneous version could start during different development stages. In Scenario 1, a manufacturer may decide to concurrently develop both the intravenous and the subcutaneous dosing alternatives up until the completion of phase 1 clinical studies and at this time point choose to only bring the subcutaneous formulation forward based on the technical feasibility of the formulation and the local tolerability in humans. Phase 2 and 3 clinical studies would subsequently be conducted with the subcutaneous formulation. In this case, only the subcutaneous product presentation would be registered.

In Scenario 2, a manufacturer would develop the intravenous formulation up until completion of phase 2 clinical trials. Subcutaneous formulation development would commence in parallel to phase 2. In Scenario 2a, the pivotal phase 3 trial would be conducted with the subcutaneous formulation, following the completion of local tolerability assessments in a nonclinical species and a pharmacokinetic noninferiority study to determine the dose that results in comparable exposure to the intravenous dosing regimen. In this scenario, only the subcutaneous formulation would be registered. In Scenario 2b, the development pathway for the subcutaneous dosing regimen would be the same as in Scenario 2a, but the manufacturer would run an additional phase 3 study with the intravenous dosing regimen to enable registration of both dosing alternatives. Notably, depending on the timing of the development programs, the phase 3 trials for the different administration routes may be combined into one single study.

In Scenario 3, development of the subcutaneous formulation would start following market entry of the intravenous version. In this case, the subcutaneous development program would be the same as in Scenarios 2a and 2b. Fig. 6.3 outlines the described scenarios for developing a subcutaneous in relation to an intravenous dosing regimen.

Development options for a subcutaneous injection device

A prerequisite for enabling convenient subcutaneous home- and self-administration is the availability of an injection device. For subcutaneous administration this could either be a PFS, an AI or a wearable OBDS. Assuming feasibility of a high-concentration subcutaneous formulation and compatibility with the primary device container, two fundamental device development options can be distinguished.

In option 1, a novel device platform needs to be developed from the beginning and may therefore not yet be available for the pivotal phase 3 study. Moreover, regulators would request the demonstration of pharmacokinetic comparability between subcutaneous injection with a handheld syringe, a PFS, or an IS in the pivotal study for the mAb and the novel automated injection device, such as an AI or a wearable OBDS, according to what is currently done for the majority of biotherapeutics (Hu et al.,

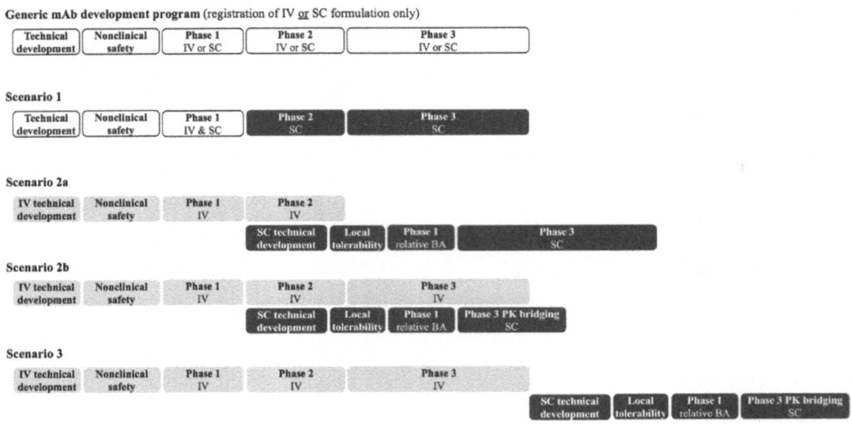

Figure 6.3 Possible development scenarios for a subcutaneous in relation to an intravenous dosing regimen. *Abbreviations: IV: intravenous; SC: subcutaneous; IV & SC: intravenous and subcutaneous; BA: bioavailability; mAb: monoclonal antibody; Scenario 1: Concurrent SC and IV development program to end Phase 1; Scenario 2a: SC formulation development started after Phase 1 (registration of SC formulation only) Scenario 2b: SC formulation development started after Phase 1 (registration of IV and SC formulations); Scenario 3: SC formulation development started after Phase 3 (registration of IV and SC formulations).*

2020). This pharmacokinetic study could start as soon as the device is clinical trial ready, likely resulting in a sequential launch of the novel biological in the presentation used in phase 3 followed by the new device as a line extension.

In the second conceivable scenario, a manufacturer could embark into the development of a device platform that is already marketed or in late-state development with a different molecule. Leveraging synergies from data generated with other molecules, such as demonstrated design verification, usability, and implemented manufacturing equipment and processes, can accelerate technical development and ideally could make the device available for the conduct of the pivotal phase 3 study and thus omit the need for conducting an additional pharmacokinetic comparability as in option 1.

Chapter 3, described a hypothetical molecule-independent device bridging approach in which an established device platform could be used without the need for generating pharmacokinetic comparability data for eligible novel molecules using the same established device platform. Provided regulators would accept this approach, also in option 1, the device platform could be available at initial molecule launch. A more detailed review on the practical considerations in clinical strategy to support the development of injectable drug-device combination products for biologics was provided previously (Li & Easton, 2018).

Summary and conclusions

It is generally recommended to consider and customize product optimizations from the start of a biological medicine's development at an ongoing basis. Manufacturers should assess the feasibility of optimizing the formulation or dosing regimen already in nonclinical pharmacokinetic and toxicology and clinical phase 1 studies. Early investment into formulation and device lifecycle management has to be conceived in connection with its expected impact on patient experience, the healthcare system, and society as a whole.

Depending on technical feasibility, the development of new administration routes, formulations or drug delivery devices can either be done within the pivotal phase 3 studies or in parallel or subsequently in dedicated studies. Clinical profiling of product optimizations within the registrational studies for a new biological might be supported by the availability of established platform technologies manufacturers can embark on. This approach avoids the need to start device development from scratch for every molecule using the same device type.

Formulation and device lifecycle management for marketed products is commonly considered on the basis of evolving real-world evidence on patient and healthcare provider needs. This stepwise product optimization approach requires a more extensive and costly development program (i.e., conduct of additional phase 1 and phase 3 studies), however at reduced risk by investing in a proven therapeutic and by having more disease insights.

Ideally, from a patient and healthcare system perspective, a biotherapeutic would be available with a variety of dosing alternatives. This provides more dosing flexibility and drug administration schemes can be selected based on individual patient preferences and capabilities, as well as on the dosing regimen of a potential combination partner.

References

Bansal, D., Malla, S., Gudala, K., & Tiwari, P. (2013). Anti-counterfeit technologies: A pharmaceutical industry perspective. *Scientia Pharmaceutica*, *81*(1), 1–13. Available from https://doi.org/10.3797/scipharm.1202-03.

Beall, R. F., & Kesselheim, A. S. (2018). Tertiary patenting on drug-device combination products in the United States. *Nature Biotechnology*, *36*(2), 142–145. Available from https://doi.org/10.1038/nbt.4078.

Berger, M. (2004). Subcutaneous immunoglobulin replacement in primary immunodeficiencies. *Clinical Immunology*, *112*(1), 1–7. Available from https://doi.org/10.1016/j.clim.2004.02.002.

Bittner, B., Chiesi, C. S., Kharawala, S., Kaur, G., & Schmidt, J. (2019). Connected drug delivery devices to complement drug treatments: Potential to facilitate disease management in home setting. *Medical Devices: Evidence and Research*, *12*, 101–127. Available from https://doi.org/10.2147/MDER.S198943.

Bittner, B., McIntyre, C., Tian, H., Tang, K., Shah, N., Phuapradit, W., Ahmed, H., Chokshi, H., Infeld, M., Fotaki, N., Ma, H., Portron, A., Jordan, P., & Schmidt, J. (2012). Phase I clinical study to select a novel oral formulation for ibandronate containing the excipient sodium N-[8-(2-hydroxybenzoyl) amino] caprylate (SNAC). *Pharmazie*, *67*(3), 233–241. Available from https://doi.org/10.1691/ph.2012.1648.

Bittner, B., Richter, W., Hourcade-Potelleret, F., McIntyre, C., Herting, F., Zepeda, M., & Schmidt, J. (2012). Development of a subcutaneous formulation for trastuzumab—Nonclinical and clinical bridging approach to the approved intravenous dosing regimen. *Arzneimittelforschung*, 401–409. Available from https://doi.org/10.1055/s-0032-1321831.

Bittner, B., Richter, W., & Schmidt, J. (2018). Subcutaneous administration of biotherapeutics: An overview of current challenges and opportunities. *BioDrugs*, *32*(5), 425–440. Available from https://doi.org/10.1007/s40259-018-0295-0.

Bruin, G., Hockey, H. U. P., La Stella, P., Sigurgeirsson, B., Fu, R., Patekar, M., Charef, P., Woessner, R., & Boutouyrie-Dumont, B. (2020). Comparison of pharmacokinetics, safety and tolerability of secukinumab administered subcutaneously using different delivery systems in healthy volunteers and in psoriasis patients. *British Journal of Clinical Pharmacology*, *86*(2), 338–351. Available from https://doi.org/10.1111/bcp.14155.

Chatelut, E., White-Koning, M. L., Mathijssen, R. H., Puisset, F., Baker, S. D., & Sparreboom, A. (2012). Dose banding as an alternative to body surface area-based dosing of chemotherapeutic agents. *British Journal of Cancer*, *107*(7), 1100–1106. Available from https://doi.org/10.1038/bjc.2012.357.

Chow, S. C., & Chang, M. (2008). Adaptive design methods in clinical trials—A review. *Orphanet Journal of Rare Diseases*, *3*(1). Available from https://doi.org/10.1186/1750-1172-3-11.

Clemens, P., Xu, S., Luo, M., Chari, A., Usmani, S., Mateos, M.-V., van de Donk, N., Kaufman, J., Moreau, P., Oriol, A., Plesner, T., Benboubker, L., San-Miguel, J., Sun, X. Y.-N., Farnsworth, A., Masterson, T., Hellemans, P., Qi, M., & Nahi, H. (2018). Pharmacokinetics (PK) of subcutaneous daratumumab in patients with relapsed or refractory (RR) multiple myeloma (MM): Primary clinical pharmacology analysis of the open-label, multicenter, phase 1b study (PAVO). *Blood*. Available from https://doi.org/10.1182/blood-2018-99-113161.

Cáceres, M. C., Guerrero-Martín, J., Pérez-Civantos, D., Palomo-López, P., Delgado-Mingorance, J. I., & Durán-Gómez, N. (2019). The importance of early identification of infusion-related reactions to monoclonal antibodies. *Therapeutics and Clinical Risk Management*, *15*, 965–977. Available from https://doi.org/10.2147/TCRM.S204909.

FDA. (2016). FDA Code of Federal Regulations 21CFR314.108. https://ecfr.federalregister.gov/.

Gardulf, A. (2007). Immunoglobulin treatment for primary antibody deficiencies: Advantages of the subcutaneous route. *BioDrugs*, *21*(2), 105–116. Available from https://doi.org/10.2165/00063030-200721020-00005.

Gupta, H., Kumar, S., Roy, S., & Gaud, R. (2010). Patent protection strategies. *Journal of Pharmacy And Bioallied Sciences*, 2. Available from https://doi.org/10.4103/0975-7406.62694.

Hu, P., Wang, J., Florian, J., Shatzer, K., Stevens, A. M., Gertz, J., Ji, P., Huang, S. M., Zineh, I., & Wang, Y. M. C. (2020). Systematic review of device parameters and design of studies bridging biologic-device combination products using prefilled syringes and autoinjectors. *AAPS Journal*, *22*(2). Available from https://doi.org/10.1208/s12248-020-0433-8.

Hui, T., Mohammed, B., Donyai, P., McCrindle, R., & Sherratt, R. (2020). Enhancing pharmaceutical packaging through a technology ecosystem to facilitate the reuse of medicines and reduce medicinal waste. *Pharmacy*, 58. Available from https://doi.org/10.3390/pharmacy8020058.

Klonoff, D. C., & Kerr, D. (2018). Smart pens will improve insulin therapy. *Journal of Diabetes Science and Technology*, *12*(3), 551−553. Available from https://doi.org/10.1177/1932296818759845.

Klonoff, D. C., Zhang, J. Y., Shang, T., Mehta, C., & Kerr, D. (2020). Pharmacoadherence: An opportunity for digital health to inform the third dimension of pharmacotherapy for diabetes. *Journal of Diabetes Science and Technology*, *15*(1), 177−183. Available from https://doi.org/10.1177/1932296820973185.

Li, Z., & Easton, R. (2018). Practical considerations in clinical strategy to support the development of injectable drug-device combination products for biologics. *MAbs*, *10*(1), 18−33. Available from https://doi.org/10.1080/19420862.2017.1392424.

Lu, R. M., Hwang, Y. C., Liu, I. J., Lee, C. C., Tsai, H. Z., Li, H. J., & Wu, H. C. (2020). Development of therapeutic antibodies for the treatment of diseases. *Journal of Biomedical Science*, *27*(1). Available from https://doi.org/10.1186/s12929-019-0592-z.

Mahan, V. (2014). Clinical trial phases. *International Journal of Clinical Medicine*, 1374−1383. Available from https://doi.org/10.4236/ijcm.2014.521175.

Paul, B., Atrash, S., Bhutani, M., Voorhees, P., Hamadeh, I., & Usmani, S. Z. (2020). An evaluation of subcutaneous daratumumab for the treatment of multiple myeloma. *Expert Review of Hematology*. Available from https://doi.org/10.1080/17474086.2020.1795829.

Pisano, M. (2014). Overview of insulin and non-insulin delivery devices in the treatment of diabetes. *P and T*, *39*(12), 866−876. Available from http://www.ptcommunity.com/system/files/pdf/ptj3912866.pdf.

Quinn, A. E., Trachtenberg, A. J., McBrien, K. A., Ogundeji, Y., Souri, S., Manns, L., Rennert-May, E., Ronksley, P., Au, F., Arora, N., Hemmelgarn, B., Tonelli, M., & Manns, B. J. (2020). Impact of payment model on the behaviour of specialist physicians: A systematic review. *Health Policy*, *124*(4), 345−358. Available from https://doi.org/10.1016/j.healthpol.2020.02.007.

Rahnfeld, L., & Luciani, P. (2020). Injectable lipid-based depot formulations: Where do we stand? *Pharmaceutics*, *12*(6), 1−28. Available from https://doi.org/10.3390/pharmaceutics12060567.

Richter, W. F., & Jacobsen, B. (2014). Subcutaneous absorption of biotherapeutics: Knowns and unknowns. *Drug Metabolism and Disposition*, *42*(11), 1881−1889. Available from https://doi.org/10.1124/dmd.114.059238.

Rombouts, M. D., Swart, E. L., Van Den Eertwegh, A. J. M., & Crul, M. (2020). Systematic review on infusion reactions to and infusion rate of monoclonal antibodies used in cancer treatment. *Anticancer Research*, *40*(3), 1201−1218. Available from https://doi.org/10.21873/anticanres.14062.

Rule, S., Collins, G. P., & Samanta, K. (2014). Subcutaneous vs intravenous rituximab in patients with non-Hodgkin lymphoma: A time and motion study in the United Kingdom. *Journal of Medical Economics*, *17*(7), 459−468. Available from https://doi.org/10.3111/13696998.2014.914033.

Sabzi, Z., Mohammadi, R., Talebi, R., & Roshandel, G. R. (2019). Medication errors and their relationship with care complexity and work dynamics. *Open Access Macedonian Journal of Medical Sciences*, 7(21), 3579–3583. Available from https://doi.org/10.3889/oamjms.2019.722.

Schaefer, E., Abbaraju, S., Walsh, M., Newman, D., Salmon, J., Amin, R., Weiss, S., Grau, U., Velagaleti, P., & Gilger, B. (2016). Sustained release of protein therapeutics from subcutaneous thermosensitive biocompatible and biodegradable pentablock copolymers (PTS gels). *Journal of Drug Delivery*, 1–15. Available from https://doi.org/10.1155/2016/2407459.

Schwendeman, S. P., Shah, R. B., Bailey, B. A., & Schwendeman, A. S. (2014). Injectable controlled release depots for large molecules. *Journal of Controlled Release*, *190*, 240–253. Available from https://doi.org/10.1016/j.jconrel.2014.05.057.

Sehn, L. H., Donaldson, J., Filewich, A., Fitzgerald, C., Gill, K. K., Runzer, N., Searle, B., Souliere, S., Spinelli, J. J., Sutherland, J., & Connors, J. M. (2007). Rapid infusion rituximab in combination with corticosteroid-containing chemotherapy or as maintenance therapy is well tolerated and can safely be delivered in the community setting. *Blood*, *109*(10), 4171–4173. Available from https://doi.org/10.1182/blood-2006-11-059469.

Shah, R., Dey, D., Pietzonka, T., Obeng, P., Ashiru, B., Schiestl, M., Cavey, A., Nkansah, E., Radiere, G., Spector, J., & Scott, C. (2021). Determinants of use of biotherapeutics in sub-Saharan Africa. *Trends in Pharmacological Sciences*, *42*(2), 75–84. Available from https://doi.org/10.1016/j.tips.2020.11.012.

Shin, W., Yang, A. Y., Yun, H., Cho, D. Y., Park, K. H., Shin, H., & Kim, A. (2020). Bioequivalence of the pharmacokinetics between tofacitinib aspartate and tofacitinib citrate in healthy subjects. *Translational and Clinical Pharmacology*, *28*(3), 160–167. Available from https://doi.org/10.12793/tcp.2020.28.e13.

Shpilberg, O., & Jackisch, C. (2013). Subcutaneous administration of rituximab (MabThera) and trastuzumab (Herceptin) using hyaluronidase. *British Journal of Cancer*, *109*(6), 1556–1561. Available from https://doi.org/10.1038/bjc.2013.371.

Singh, G. (2017). *Data exclusivity. Pharmaceutical Medicine and Translational Clinical Research* (pp. 313–323). Elsevier Inc. Available from https://doi.org/10.1016/B978-0-12-802103-3.00020-1.

Sykes, C. (2018). Time- and temperature-controlled transport: Supply chain challenges and solutions. *P and T*, *43*(3), 154–170. Available from https://www.ptcommunity.com/system/files/pdf/ptj4303154.pdf.

Tjalma, W. A. A., Van den Mooter, T., Mertens, T., Bastiaens, V., Huizing, M. T., & Papadimitriou, K. (2018). Subcutaneous trastuzumab (Herceptin) versus intravenous trastuzumab for the treatment of patients with HER2-positive breast cancer: A time, motion and cost assessment study in a lean operating day care oncology unit. *European Journal of Obstetrics and Gynecology and Reproductive Biology*, *221*, 46–51. Available from https://doi.org/10.1016/j.ejogrb.2017.12.006.

Vlaanderen, F. P., Tanke, M. A., Bloem, B. R., Faber, M. J., Eijkenaar, F., Schut, F. T., & Jeurissen, P. P. T. (2019). Design and effects of outcome-based payment models in healthcare: A systematic review. *European Journal of Health Economics*, *20*(2), 217–232. Available from https://doi.org/10.1007/s10198-018-0989-8.

Vogel, W. H. (2010). Infusion reactions: Diagnosis, assessment, and management. *Clinical Journal of Oncology Nursing*, *14*(2), E10–E21. Available from https://doi.org/10.1188/10.CJON.E10-E21.

CHAPTER 7

Summary and outlook

Summary

Formulation and device lifecycle management of biotherapeutics aims at improving the drug delivery profile and product presentation of a biological medicine. The ultimate goal is to reduce the dosing complexity of parenteral administration and, thus enhance patient experience and drug administration-related healthcare capacity. This book is meant as a repository for scientists, researchers, and drug development and commercialization professionals to get insights into the current focus of drug delivery technologies in providing patient-centric product presentations, reducing the overall treatment burden, and complementing combination therapy with one or more biological medicines dosed concomitantly.

Chapter 1 is a general review of established and emerging formulation and device lifecycle management across various disease areas and administration routes. Established product optimization activities intend at improving the product presentation and include efforts to extend stability and handling times of a formulation or provide easy to use and to store packaging options that at the same time avoid counterfeit drugs by applying secure and nonclonable technology.

As parenteral drug delivery can pose a burden to patients and healthcare providers, efforts are emerging to simplify complex drug administration. Related developments include fast intravenous infusion regimens, less-invasive subcutaneous formulations, automated and ready-to-use autoinjector (AI) devices or wearable on-body delivery systems (OBDSs), possibly equipped with connected features, and accompanying health applications. Today, these so-called smart devices are being broadly evaluated to ensure adherence to treatment in a noncontrolled setting. This includes tools, such as adherence trackers, personalized dosing reminders, and electronic features, securing a remote dialog between a patient and the treating physician while drug administration takes place in the patient's home.

Pertaining to parenteral combination therapy, product optimizations early on have to focus on harmonizing the dosing regimens between

Formulation and Device Lifecycle Management of Biotherapeutics
DOI: https://doi.org/10.1016/B978-0-12-823741-0.00008-7
© 2022 Elsevier Inc.
All rights reserved.

combination partners. In the event that one biological combination partner is dosed intravenously and the other one subcutaneously, efforts should be made to provide subcutaneous dosing alternatives for both medicines. Also, if permitted by the intrinsic half-lives of the two products, the dosing frequency should be aligned to avoid too many drug administration days and associated in-clinic visits. Ideally, the combination partners should be made available as a fixed-dose combination in addition to the individual single-component products. This approach would allow for combination with other products, while at the same time reducing the dosing complexity of the particular combination.

From the alternative administration routes reviewed in Chapter 1, intravitreal and brain delivery pose the greatest challenges to drug developers due to the relative invasiveness of the dosing procedure. Research in the field focuses on applying long-acting drug delivery technologies that have the potential to reduce the dosing frequency and, thus, lessen the drug administration-related burden. Today, delivering biological agents to the brain would typically require parenteral high-dose formulations due to the typically low blood–brain barrier penetration. Consequently, invasive implants or intrathecal injections are currently part of the standard of care. Here, the combination of different formulation and device technologies may facilitate improved and targeted delivery for future dosing. For example, combining high-concentration technologies that reduce the overall volume of the dosing solution with nanoparticular or micellar agents that enhance blood–brain barrier penetration may enable efficacy at lower dose levels and dosing volumes. Administration of these formulations by means of ready-to-use subcutaneous injection devices may have the potential to further increase patient experience with the dosing regimen.

Chapter 2 reviewed the nonclinical and clinical development pathway for subcutaneous dosing regimens as an alternative to established intravenous treatments for monoclonal antibodies (mAbs). The underlying clinical bridging approach assumed that to achieve the same degree of target receptor saturation and, thus, comparable efficacy outcomes with both administration routes, comparable exposure [expressed as area under the serum concentration-time curve (AUC) and/or serum trough concentrations (C_{trough})] needs to be reached.

Notably, the regulatory requirements for a change from immediate to slow release oral formulations for small molecules did serve as a precedent for the development program of the novel subcutaneous formulations for mAbs. As oral delivery of small molecules with immediate release and slow release formulation technologies results in different pharmacokinetic profiles, in such situations, demonstrating equivalent efficacy via a bioequivalence study is not feasible. Instead, bridging from an immediate release to a so-called long-acting formulation has to be based on an established pharmacokinetic-efficacy correlation and can involve the generation of additional comparative clinical data. Different to small molecules, due to the absorption through the lymphatic system following subcutaneous dosing, immunogenicity is a key additional parameter in clinical trials supporting the registration of an administration route change for mAbs.

When initiating the development program for subcutaneous trastuzumab, the first subcutaneous high-volume formulation in oncology, there was a scientific debate on whether maximum serum levels (C_{max}) of the antibody would be relevant for efficacy outcomes. If this had been the case, the described pharmacokinetic-based bridging approach would not have been feasible. The advantage of starting with trastuzumab as the first mAb applying the pharmacokinetic-based bridging approach outlined above was that the intravenous formulation was on the market with two different dosing regimens. Namely, a weekly (q1w) regimen and an every 3 weeks (q3w) regimen were approved, resulting in comparable efficacy. Both intravenous dosing regimens yielded a comparable AUC and exceeded the C_{trough} threshold of 20 μg/mL. However, inherent to the different individual dose levels with q1w and q3w dosing, C_{max} values were markedly different. This observation supported the hypothesis that C_{max} would not drive efficacy outcomes.

Consequently, it was hypothesized that a subcutaneous formulation that results in noninferior exposure to the intravenous dosing regimens would result in comparable efficacy. This assumption was confirmed with the results of a phase 3 noninferiority study. This pharmacokinetic-based clinical bridging program was complemented with technical data, demonstrating that a change in the composition of the dosing solution does not impact molecular integrity of the mAb. Another prerequisite for applying this bridging approach was that the dosing frequency with the subcutaneous trastuzumab formulation remained the same as with intravenous infusion.

The pharmacokinetic-based bridging approach is exemplified in the following for trastuzumab. Here, the development pathway started with nonclinical toxicology and tolerability studies to confirm that the change to subcutaneous dosing and the resulting absorption via the lymphatics would not impair the established efficacy and safety profile of trastuzumab with intravenous dosing. In addition, a nonclinical xenograft study assessing the impact of the altered pharmacokinetic profile on the pharmacodynamic effects revealed that the overall exposure and/or C_{trough} rather than C_{max} drives efficacy outcomes.

The clinical bridging program commenced with an adaptive dose finding/dose confirmation trial. Within the same study, the subcutaneous dose expected to result in noninferior C_{trough} and AUC to intravenous dosing was initially calculated based on the pharmacokinetic data from healthy volunteers and subsequently confirmed in patients with HER2-positive early breast cancer. Subcutaneous dosing in this phase 1b study was still conducted with a body weight-adjusted dose, the approved regimen for the intravenous formulation. A modeling and simulation approach utilizing the established pharmacokinetic data derived from the intravenous regimen was subsequently applied to predict a fixed subcutaneous dose, avoiding the need to determine an individual dose for each patient. In addition, the subcutaneous dose was calculated to omit the loading dose, required for the intravenous regimen, while ensuring comparable C_{trough} is reached from cycle 1 onward. This subcutaneous dosing regimen was then compared to the intravenous regimen in a noninferiority study in patients with HER2-positive eBC in the neoadjuvant setting with C_{trough} at predose cycle 8 and pathological complete response as coprimary endpoints.

Today, this bridging approach from intravenous to subcutaneous dosing of high-volume mAbs is broadly applied across oncology indications and supported the approval of trastuzumab and the fixed dose combination of pertuzumab and trastuzumab in HER2-positive breast cancer, rituximab in malignant lymphoma, and daratumumab in multiple myeloma. It is expected that the same scientific concepts can also be applied to mAbs in other disease areas, naturally considering mAb- and indication-specific requirements.

With these high-volume subcutaneous dosing regimens made available in oncology, a first step in reducing drug administration-related costs and resources was made. While dosing still takes place in a controlled healthcare institutional setting, patients and healthcare providers acknowledge the reduced treatment time that frees up resources and lessens the overall burden inherent to parenteral drug administration. As a next step, home administration of subcutaneous biotherapeutics should be realized for eligible treatments and patients in cancer care. An essential condition for convenient, safe, and reliable high-volume subcutaneous dosing outside of a controlled in-clinic environment is the availability of ready-to-use wearable drug delivery devices. The first marketed high-volume patch device delivers 3.5 mL of evolocumab, a lipid-lowering mAb. To account for the even higher dosing volumes for the subcutaneous mAb formulations in oncology (between 5 and 15 mL), further development work is required to provide adequate OBDSs. Especially with the onset of the COVID-19 pandemic, embarking into attempts to enable home- and possibly patient self-administration has turned into a necessity for the healthcare system.

To assess the next steps in accelerating the underlying device development pathways while still generating a scientifically robust data package, Chapter 3 discussed the current requirements for validating automated injection devices clinically. Focus is on AIs for injection volumes of up to approximately 2 mL and OBDSs for higher volumes of up to approximately 50 mL. While customized and ready-to-use subcutaneous AI device platforms have become indispensable in many disease areas, OBDS devices are a relative novelty in the space.

The challenge inherent to applying such novel device technologies is that respective biological device combinations are typically not available for inclusion into the pivotal clinical program that supports initial approval of a novel biological agent. In such occasions, in the phase 3 trials, the subcutaneous formulation is commonly supplied with a prefilled syringe (PFS) for lower dosing volumes or in a vial presentation for higher volumes. The containing dosing solution is administered either manually by means of a handheld syringe (HHS) or semiautomatedly using an infusion set (IS) and a syringe pump.

Therefore, dosing in a flexible care setting would in many cases only be feasible as a second step and not yet with first market entry of the molecule. In addition to the required technical data package, manufacturers have to demonstrate clinically that the automated injection device is easy to use, thus allowing application in the intended use environment. Especially in case a device technology is foreseen for home- and self-administration, usability studies that confirm safe and efficacious use as well as intuitiveness of the instruction for use materials are a critical integral element of the overall data package. These studies are most frequently conducted as simulated use trials in which injection by patients, lay caregivers, or healthcare professionals is performed into a pad representing the subcutaneous injection site. In some occasions, so-called "real use" or "actual use" studies can be mandated depending on the anticipated risk profile with the usage of the device related to the safety and tolerability profile of the active molecule and the expected capability of the patient and caregiver population to use the device without professional supervision. On top of these human factors studies, pharmacokinetic comparability studies are conducted to assess whether a change from manual injection with a PFS or HHS or semiautomated injection with an IS in the pivotal trial versus using an automated AI or OBDS technology would result in a clinically relevant change in the pharmacokinetic profile of the molecule.

The conduct of such pharmacokinetic comparability trials for each individual mAb using the same automated device platform adds to the overall development cost and timelines and, depending on the safety and tolerability profile of a mAb, may have to be conducted in patients. Recruiting patients into pharmacokinetic studies, with often only one or two doses administered, can pose a challenge from an operational perspective and even further delay the option for dosing outside of the clinic.

Chapter 3 proposed an alternative molecule-independent device bridging approach that would require the conduct of only one pharmacokinetic comparability study per automated AI or OBDS platform used across several mAbs. The device would be qualified within its operational volume and viscosity ranges using a "model mAb." This "model mAb" would be representative of other mAbs with similar physicochemical and pharmacokinetic properties and would have previously been demonstrated to be safely dosed to healthy volunteers.

The underlying hypothesis is that with an automated AI or OBDS technology, the injection procedure would be standardized compared to the more variable administration conditions with manual or semiautomated injection in the pivotal clinical trial. Moreover, differences in the particular injection procedure are expected to only affect the absorption phase of a mAb, while the systemic distribution and elimination profile would be unchanged. Eligible mAbs would possess a slow absorption into the systemic circulation, so that any change in the administration time would be negligible. A prerequisite for this molecule-independent device bridging approach is that the same formulation at the same dosing volume is used in the pivotal study and in the automated device. Also, the injection procedures across the study sites in the pivotal trial have to be harmonized with respect to the injection depth, angle, and infusion rate to reflect the operational range of the envisaged automated device platform. It is of note that the proposed conceptual molecule-independent device bridging approach requires further consensus building in the scientific community and with global regulatory agencies.

A biological product that is safe, tolerated, and straightforward to use is a critical, but not the only enabler for a flexible care setting in cancer care. As outlined in Chapter 4, considering the complexity and invasiveness of the parenteral dosing procedure, a close cooperation between manufacturers, regulators, and medical practitioners is mandated to implement measures for selecting the most appropriate patient population adept and willing to receive the parenteral treatment in a nonsupervised setting. To be capable of administering the medication independently or with the help of a lay caregiver, patients have to acknowledge and understand the drug administration procedure. Oncologists, nurses, and supporting home care service providers require appropriate training and have to be equipped and reimbursed for providing adequate professional drug administration monitoring support. Upfront investment is warranted to create an infrastructure for the delivery, dosing, and disposal of the medication, so that dosing outside of a healthcare institution can be realized without increasing the burden for patients and caregivers associated with tasks accompanying the drug administration procedure.

Precedent research on biological medicines with subcutaneous presentations that are approved for self-administration in other disease areas reveals that actual dosing volumes range from less than 1 mL as exemplified by interferons in multiple sclerosis, to up to 600 mL for human immunoglobulins in immunodeficiency indications. In the ideal case, low volume formulations are administered with automated AI devices with a fixed dosing regimen and no requirement for dose adaptation throughout the treatment period. More complex regimens, however, did not preclude acceptance for self-administration. For example, a number of precedents are dosed with a weight-adjusted regimen, or alternatively, individual efficacy outcomes or safety findings necessitate dose adjustments at an ongoing basis. Moreover, even product presentations that require more complex preparation procedures, such as reconstitution of a lyophilized powder or the manual or semiautomated preinjection of a dispersion enhancer, are found among the self-administered products.

Most subcutaneous self-administration examples are allowed for injection into more than one area of the body, so patients can administer to the abdomen, thigh, or upper arm, as preferred, and also have the option to rotate the injection site to avoid local reactions to the extent possible. As hypersensitivity or anaphylactic reactions are observed for a number of medications with self-administration in the label, dosing is generally initiated by an experienced physician in the given indication. In case the treatment is tolerated during the initial treatment cycles, eligible patients can be allowed to continue dosing at home without professional supervision. The decision on the most appropriate care setting is at the discretion of the physician, follows thorough training, and frequently includes an observation phase in which the patient or caregiver conducts the injection under supervision of a healthcare provider.

Chapters 2, 3, and 4 described the underlying rationale, the scientific conduct, as well as the societal relevance of formulation and device lifecycle management for biotherapeutics. Specifically, the reader learned about the importance of simplifying parenteral administration of biological medicines and the supporting nonclinical and clinical development pathways. The central themes were the development of high-dose subcutaneous formulations for mAbs as an alternative to more invasive intravenous infusions and the essential next steps to enabling parenteral dosing in a flexible care setting in cancer care.

It is of utmost importance that scientists, researchers, and development and commercialization experts involved in formulation and device lifecycle management have a sound understanding on when to engage into product optimizations and what available insights can be leveraged across molecules and disease areas. Starting activities from scratch without considering available knowledge will markedly add to the overall development costs and timelines, an aspect that is highly relevant, especially in light of the growing resource constraints in the healthcare sector.

Chapters 5 and 6 therefore, digested the content laid out in the previous chapters and provided recommendations on how to base new formulation or device developments on existing synergies between products that apply the same formulation and device technology. In addition, a decision architecture is provided, detailing at what stage during the lifecycle of a biological medicine manufacturers should best embark into related investments.

Synergies exist, for instance, in the design of the nonclinical and clinical development pathways and the impact of a novel drug delivery modality on the safety, tolerability, and efficacy profile of a molecule. Also, when assessing patient and healthcare provider needs and preferences for one over other product presentations, or the impact of a new technology on healthcare costs and resources, learnings from other disease areas can be broadly applied to new therapeutic areas applying the same formulation or device technology. The consequences of more convenient products for global access to medications based on local differences in healthcare provider reimbursement models, are expected to be comparable between different indications.

During the lifecycle of a biological therapy, knowledge is increasingly becoming available both on the probability of launch of a molecule as well as of the technical feasibility of a formulation or device technology. For example, phase 1 clinical trials provide initial evidence on the final dose level of the drug. Insights will inform selection of an appropriate drug delivery device based on the expected fill volume and viscosity of the formulation. This understanding derisks selection of the device type and directs the next steps to including the technology into clinical trials. Once a molecule is on the market, feedback from patients, caregivers, or providers on the product's usability profile will guide improvements of the initial product presentation and drug delivery profile. At this stage, manufacturers are typically more confident in taking larger technology investments, as the molecule has been demonstrated to be safe and efficacious and is an accepted treatment modality on the market. Any expenditure in improving usability is expected to be valued by customers.

Readers were provided with details on the evolving knowledge base from nonclinical to clinical development up until and following market entry. This decision architecture is expected to be used as a frame for researchers and drug developers for taking appropriate need-based investment decisions aiming at improving convenience and reliability of the products on an ongoing basis.

Discussion

At present, the approach to enable a flexible care setting for mAbs in cancer care is a major element of drug delivery improvements. As opposed to other disease areas, this is still a comparatively new, but rapidly emerging focus area of researchers, healthcare providers, and regulators. The main barrier toward parenteral dosing outside of the clinical setting is infusion-related reactions occurring predominantly during initial treatment cycles and, thus, frequently mandating dosing in healthcare institutions with close proximity to resuscitation facilities. In addition, high dosing volumes for mAbs pose a challenge with regard to developing convenient and ready-to-use subcutaneous formulations.

Inherent to any attempt in introducing new dosing strategies, it is important to be clear in articulating the challenges associated with implementing the required changes to the customers. In the opinion of the authors, a key accomplishment in the context of raising awareness of the relevance of reducing the dosing complexity of parenteral administration of biotherapeutics in cancer care was to discuss the concept of enabling dosing in a "flexible care setting". Provided the drug has been shown to be safe and tolerated, in such a setting, at the discretion of the treating physician patients can choose the place of drug administration based on individual preferences and capabilities. This could include specialized clinics, infusion centers, community centers located closer to the patient's home, general practitioners' offices, as well as home- and possibly even patient self-administration. Different from the term "at-home administration" that we used initially when discussing opportunities and barriers of formulation and device lifecycle management for biotherapeutics in oncology, the term "flexible care setting" in itself highlights that drug administration may take place in different locations. The term reflects that the treatment setting is truly dependent on the molecule's safety profile, patient choice, as well as on whether the patient receives comedications or examinations that need to be conducted in the clinic anyway. With the understanding that in-clinic versus at-home administration are not mutually exclusive options, but that a variety of conceivable drug administration settings exist, multidisciplinary stakeholders were increasingly aligning with the overall pathway toward patient-centric treatment management via a reduced drug administration complexity.

Another observation made during clinical development of subcutaneous dosing alternatives in oncology was that while there is significant knowledge exchange on drug delivery alternatives happening within a given therapeutic area, that is, via scientific conferences, in clinical practice, or via scientific publications in the field, cross-disease area cooperation is generally limited.

For example, the first high-volume subcutaneous administration of immune human serum globulin was reported back in 1952 when a subcutaneous infusion of approximately 20 mL was administered to an 8-year-old patient with agammaglobulinemia. Since then, a variety of immunoglobulin products in the same volume range have been approved, including for subcutaneous self-administration. Thus, even though there is significant experience in the field of high-volume subcutaneous administrations available since decades, until the introduction of high-volume mAbs in oncology, there was still the general perception by scientists and healthcare providers that subcutaneous injection of volumes greater than approximately 2 mL would not be feasible without significant pain and discomfort.

One has to acknowledge in this context that technology has evolved over time and that the feasibility of high-concentration dosing solutions and the coadministration of a dispersion enhancer have largely facilitated subcutaneous injection of high-dose biotherapeutics. This way, the injection volumes could be reduced and transient and reversible enzymatic cleavage of hyaluronan, a component of the subcutaneous tissue layer, facilitated spreading of injected larger volumes, thus reducing induration at the injection site.

It is expected that in response to the growing demand to reduce the cost and resource pressure on the healthcare system and in light of managing a pandemic, innovation in formulation and device lifecycle management will further accelerate. Subcutaneous (connected) drug delivery device technology platforms will support dosing in a flexible care setting. Homecare service providers will be broadly equipped to facilitate parenteral administration in the patient's home. In cancer care, global reimbursement systems may to some extent change from fee-for-service payments to value-based models that account for the efficiency rather than for complexity of care. Learnings from the current experience with formulating high-dose mAbs are expected to be translatable to other disease areas with similar approaches in earlier development.

From a technology perspective, methodologies that allow for the oral delivery of biopharmaceuticals and either target the molecule to a specific site in the gastrointestinal tract for local action or even result in clinically relevant systemic exposure have the potential to change the drug delivery paradigm for biotherapeutics. Given that these novel technologies are still in their infancy, the impact on the achievable dosing regimen and the overall safety, and especially the immunogenicity profile of an embedded biological agent, requires further understanding. Efforts to generate the required data are made in order not to miss an opportunity to markedly improve convenience for patients as opposed to more invasive parenteral administration.

Classically, the term "formulation and device lifecycle management" implies that the topic is predominantly about technical development activities. While such galenical work is an important building block for successfully implementing product optimizations, other disciplines play a pivotal role in realizing meaningful improvements in drug delivery. Multidisciplinary project teams, in order to make informed decisions on the most appropriate product optimization activity, require a sound understanding of market dynamics, including patient and healthcare provider needs, of nonclinical and clinical development requirements, as well as of the overall access and reimbursement situation in a given legislation. At the same time, awareness of emerging technology opportunities that could be applied to address identified needs is warranted.

The discipline of formulation and device lifecycle management of biotherapeutics and the supporting multifunctional development and commercialization activities are currently enjoying great attention and are increasingly becoming an integral part of drug development programs. In the opinion of the authors, a major future action is to foster a cross-disciplinary and cross-indication cooperation in the field to generate a fruitful experience exchange. This has to happen broadly and in close cooperation between pharmaceutical industry, academia, as well as regulatory and health technology assessment agencies.

Index

Note: Page numbers followed by "*t*" refer to tables.